Codes for Error Control and Synchronization

The Artech House Communication and Electronic Defense Library

Arbenz, Kurt, and Jean-Claude Martin, **Mathematical Methods of Information Transmission**

Arbenz, Kurt, and Alfred Wohlhauser, **Advanced Mathematics for Practicing Engineers**

Deavours, Cipher A., and Louis Kruh, **Machine Cryptography and Modern Cryptanalysis**

Deavours, Cipher A., et al., eds., **Cryptology Yesterday, Today, and Tomorrow**

Fielding, John E., and Gary D. Reynolds, **RGCALC: Radar Range Detection Software and User's Manual**

Rubin, Olis, **Analysis and Synthesis of Logic Systems**

Schleher, D. Curtis, **Introduction to Electronic Warfare**

Torrieri, Don J., **Principles of Secure Communication Systems**

Wiggert, Djimitri, **Codes for Error Control and Synchronization**

Wiley, Richard G., **Electronic Intelligence: The Analysis of Radar Signals**

Wiley, Richard G., **Electronic Intelligence: The Interception of Radar Signals**

Wiley, Richard G., and Michael B. Szymanski, **Pulse Train Analysis Using Personal Computers**

Codes for Error Control and Synchronization

Djimitri Wiggert

ARTECH HOUSE, INC.
685 Canton Street
Norwood, MA 02062

International Standard Book Number: 0-89006-166-1
Library of Congress Catalog Card Number: 88-6339

10 9 8 7 6 5 4 3 2 1

Contents

To Barbara, Lara, Thea, Heidi and Kristian

Preface

This book is both an extension and a revision of the book *Error-Control Coding and Applications* first published by Artech House in 1978. The current title reflects the addition (as Chapter 12) of introductory material on synchronization codes. Other new material includes:

- a discussion and application of the interrelationship and trade-off of capabilities for simultaneous detection and correction of errors with block codes;
- a more extensive discussion of decoding for BCH codes;
- additional material on Reed-Solomon codes;
- a discussion of nonbinary convolutional codes;
- performance plots of convolutional codes with threshold decoding;
- a qualitative discussion of soft decisions.

It is a pleasure to acknowledge the contributions of two very special people: Ms. Catherine A. Smith for her excellent typing support, and Mr. Robert T. Garrow for his unsurpassed skill in producing professional quality graphics. I am also grateful to the MITRE Corportion for making the support of these two individuals available to me.

Chapter 1
Introduction and Overview

1.1 GENERAL

This book examines error-control techniques (involving error detection as well as correction) from the point of view of the user; that is, the individual who needs to cope with errors in a specific instance of data transmission or data storage. The presentation is also aimed at those who evaluate work done for the proposal, implementation, and testing of an error-control system, as well as those who compare two or more systems. In most cases, we will present results without proof or derivation. However, we will usually discuss why the result is useful or significant. Wherever it seems helpful to do so, plausibility arguments will be given. Examples will be introduced at many points along the way as illustrations of the theory. In fact, the same example may be brought through several stages by being used in conjunction with several different points of the theory.

A person who is considering the use of error control should ask the following questions:

1. How much improvement in performance is possible with error control?
2. How easily and at what cost can error control be implemented?
3. Is error control alone the best solution to a problem, or should it be supplemented (or even replaced) by another technique?

The first question requires as complete a knowledge as possible of the noise or error statistics of the unprotected transmission or storage system so that we can predict performance of various types of error control. A measure of effectiveness used to help answer question 1 might be one of the following: decrease in error probability for fixed signal-to-noise ratio; decrease in transmitted power or antenna gain for a fixed probability of error; the length of the maximum correctable burst of errors.

The second question addresses the following engineering problems: cost, equipment, and software complexity; speed required *versus* speed available; overall throughput rate; trade-offs against other coding and noncoding error-control techniques. From the system engineering point of view, this question is as important as the first. If you can obtain adequate error protection through error-control coding only by delaying final system output until it is no longer useful or by doubling the cost of the system without error control, for example, then you should certainly consider other methods.

The preceding discussion leads directly to the third question. The engineer should keep an open mind with regard to alternative techniques. For instance, it may prove less costly and sufficiently effective, compared with error control, to employ some form of space or frequency diversity or to take advantage of spectrum-spreading techniques (particularly if the latter techniques are already being contemplated for their antijamming or data security properties).

1.2 OVERVIEW OF THE BOOK

This introductory chapter will conclude with a presentation and discussion of a model of a digital communication (or storage) system. Chapter 2 presents some terminology and analysis that connect communication waveforms and the discrete entities (digits) used in error-control coding.

Chapter 3 introduces the notions of linear algebra as the basis for decoding techniques for parity-check block codes, the most easily understood of all codes. The important relationship between distance and error-control properties is introduced in this chapter. Chapter 4 develops enough of the results of abstract algebra (particularly groups and finite fields) and cyclic code theory so that we can understand and appreciate the two classes of cyclic codes: the Bose-Chaudhuri-Hocquenghem (BCH) and Reed-Solomon codes, which are discussed in Chapter 5. The BCH codes are the best block codes in general use for dealing with independent errors, while the Reed-Solomon codes are best for coping with bursts of errors.

Chapters 6 through 9 are devoted to convolutional codes—their properties, representation, and encoding—and to the three leading decoding techniques. Threshold decoding, sequential decoding, and maximum-likelihood (Viterbi) decoding are all very effective techniques that excel under different circumstances.

Burst-correcting techniques, including the use of error detection and retransmission, are the subject of Chapter 10. Chapter 11 provides a useful discussion of the trade-offs involved in selecting a coding scheme for each of several real-life situations in which the channel statistics are known, or at least strongly suspected, with emphasis on the engineering aspects of the problem. Chapter 12 presents synchronization codes, and the concluding appendix provides an elementary, detailed development of vectors and matrices.

1.3 MODEL OF A COMMUNICATION SYSTEM

Several useful models of a communication system can be used as a basis for discussion of communication and coding. A simple model that emphasizes the role of error control is shown in Figure 1.1. This model consists of a source, an error-control encoder, a noisy channel, an error-control decoder, and a sink or user, which are connected in cascade as shown. For this system, all information transferred between system blocks must be in digital form (usually 1s and 0s) in order for error control to be applied in the manner shown. It is common practice to expand the source, channel, and sink blocks to produce the model shown in Figure 1.2. The dashed lines relate these new components of the model to those in Figure 1.1.

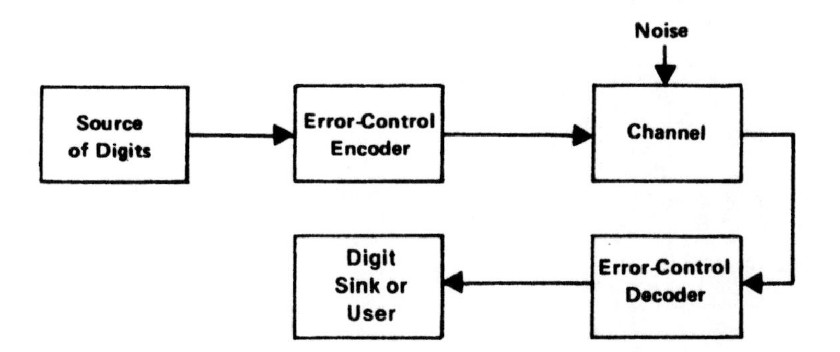

Figure 1.1 Simple model of a communication or data storage and retrieval system.

The message generator may have either a digital (discrete amplitude) or analog (continuous amplitude) output. To give an idea of the diverse quantities that could be considered messages, we mention the following examples of digital messages:

- Final score of a football game;
- Dow Jones closing average;

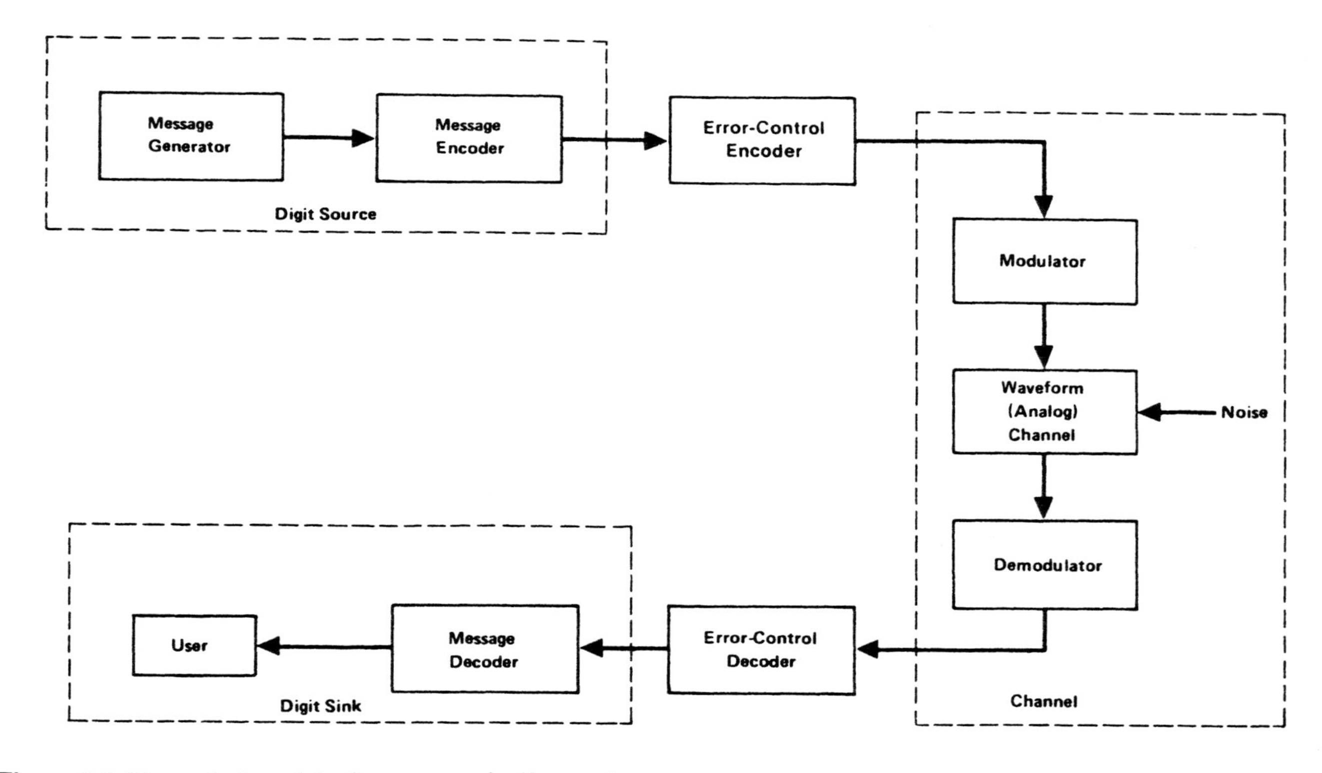

Figure 1.2 Expanded model of a communication system.

- Amount of credit card purchase;
- Number, kind, position, and velocity of vehicles spotted by an airborne observer.

In contrast, an example of an analog message is the real-time record of the blood pressure of an orbiting astronaut.

Because the output of the message encoder must be suitable for input to the error-control encoder, this output must be digital and is permitted to change only at specified times. Thus, if the message generator produces analog output, the message encoder must as a minimum possess analog-to-digital capability—for example, pulse code modulation (PCM). The message encoder may also remove redundancy from the message. For example, an astronaut's blood pressure will be essentially constant for long periods of time. Thus, time and bandwidth need not transmit each and every value, either sampled or continuous. In this case, it may be best to transmit (e.g., by run-length encoding) only the changes exceeding some preset threshold. Finally, if the error-control encoder operates with a binary alphabet, you may need to convert in the message encoder from some larger alphabet (e.g., number of PCM quantization levels) to a sequence of 0s and 1s.

From the discussion so far, you should already be aware of the considerable variation in degree of detail that must be represented in any given situation and the considerable latitude of the person modeling the communication system. For example, if you want to encode a message for communication security purposes, you could do so at the analog message stage (e.g., spectrum scrambling) or at any of several points in the digital portion of the system.

Once the message has been digitized and redundancy removed (if desired) you should reintroduce redundancy in the error-control encoder. This time, however, the redundancy is added at a fixed rate and in a known way, which introduces structure to the error-control encoder output in the form of block codes, convolutional codes, or a combination. This output now passes into the channel. In most cases, this involves modulating the sequence of pulses onto a carrier, which is done by the block labeled "modulator."

The analog channel inevitably modifies the transmitted waveform. This modification could occur in a variety of ways, including adding noise and fading. Additive noise, such as that in a deep space channel, frequently has a Gaussian amplitude distribution and a frequency spectrum that is flat, at least over the signal bandwidth. Additive noise could also occur as impulse noise, which has a broad frequency spectrum but strongly non-Gaussian amplitude statistics. Fading of signal strength is often adequately modeled by the Rayleigh distribution. One example of a mechanism causing fading is the multipath that frequently exists between an airborne or ground transmitter and a satellite. Fading results in bursts of errors in the input, and possibly the output, of the error-control decoder.

The function of the demodulator block of Figure 1.2 includes not only recovery of the modulating waveform (now corrupted by noise), but the generation of estimates of the digits transmitted. These estimates could simply be 0s and 1s (i.e., hard decisions). Alternatively, they could be quantized estimates (soft decisions) of these digits, with the demodulator either generating values between or beyond the values corresponding to 0 and 1 or generating 0s and 1s plus a rough measure of the reliability of each digit estimate.

The error-control decoder then operates on these estimates of the transmitted digits to produce its estimate of the transmitted encoded message sequence. This decoder could be a look-up table based on a calculation of the most likely transmitted block code word, or it could be a sequential decoder, to cite only two possibilities. The estimate of the encoded transmitted message sequence is now presented to the message decoder, which in turn generates its estimate of the transmitted message.

We should emphasize that the communication system model just presented is neither unique nor necessarily the most suitable one to use in all situations. Figure 1.3 illustrates the type of complex system that we might encounter in practice. Here we have multiplexing of data, multiple levels of error control, encryption for security, and spectrum spreading for antijamming, antimultipath, multiple access, or additional security. Such a system typifies a modern civil government, large business, or military communication system carrying a mix of voice, data, and facsimile. (For an analysis of the architecture and functioning of another model of such a system, see the paper by Glenn [1.1]). The mux (multiplex) and demux functions could be carried out to place several telephone conversations (mux1) or voice and data streams (mux2) on a single channel by either time- or frequency-division multiplexing. The multiple levels of error-control coding could be a concatenated scheme using a convolutional inner code with Viterbi decoding (error control coder-decoder 2) and a block Reed-Solomon outer code (error control coder-decoder 1), in which symbols of the Reed-Solomon code become the binary digit stream entering and leaving the convolutional coder-decoder combination. Here each coder-decoder pair is matched to the channel that it serves. Thus, the convolutional code could be coping with random-error Channel 2, making Channel 1 a generally noiseless channel with occasional bursts of Viterbi decoder output errors. This latter channel is amenable to error control by the burst-error oriented Reed-Solomon code.

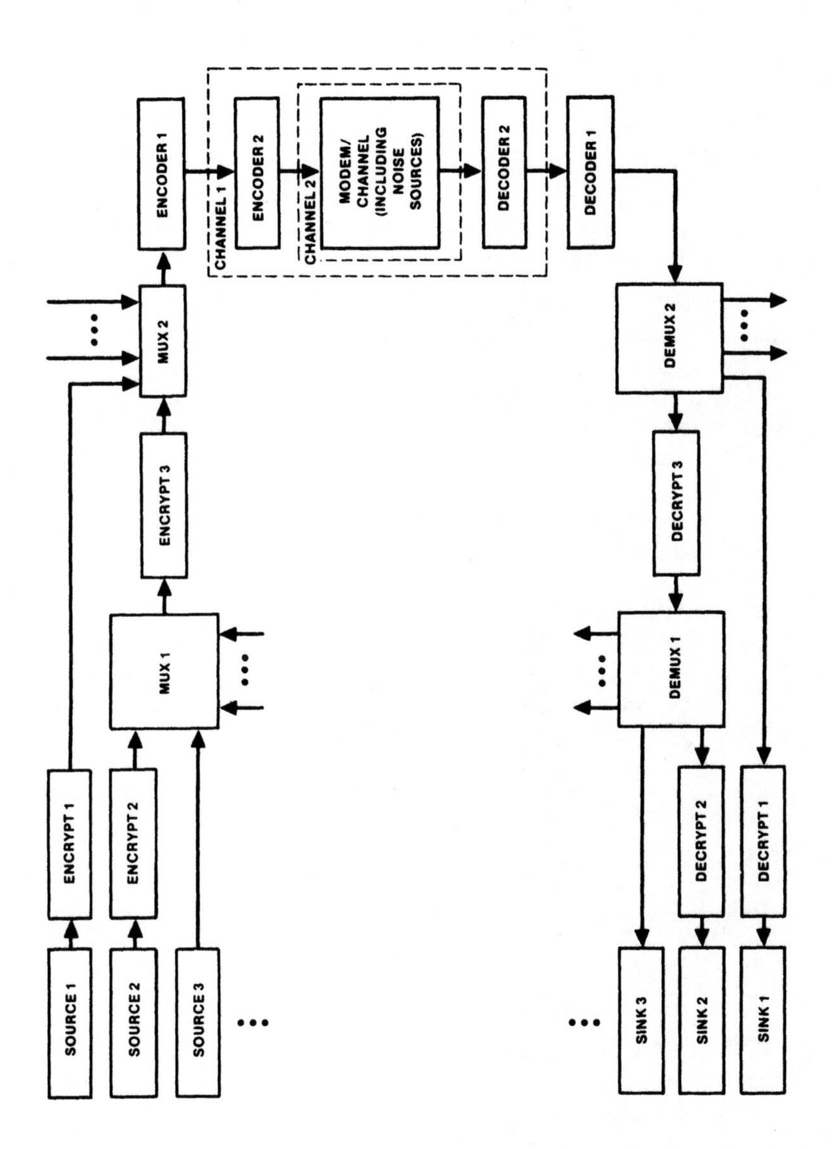

Figure 1.3 A possible "real-life" communication and data-transmission system.

Encryption, long a part of the military communication picture, has become increasingly common in the civilian sector as well, as corporations and law enforcement agencies strive to protect themselves against unwanted electronic eavesdropping.

One final comment may be useful at this point, especially to readers familiar with the seven-layer open system interconnection (OSI) model of a data communication system being promoted by the International Standards Organization (ISO). We can draw the following analogies between the communication system models shown in Figures 1.2 and 1.3 and the ISO model (see, for example, Tanenbaum [1.2]):

- Message generator and user are the application layer.
- Message encoder-decoder and encrypt-decrypt are within the presentation layer.
- Mux-demux would probably be part of the network layer.
- Error-control encoder-decoder and modulator-demodulator (modem) are all within the data link layer.
- Waveform channel is within the physical layer.
- Nothing is shown that can be clearly identified with either the transport layer or session layer.

Chapter 2
Communication Theory Considerations

2.1 INTRODUCTION

This chapter relates error control and digital communication, in which operations for the most part are performed on discrete values of variables, to waveforms, which are basically analog. *Analog* here refers to being able to assume any of a continuum of values that cover either a finite or an infinite range. Clearly, channel noise is an analog waveform although its effects can be represented digitally in terms of the output of a demodulator. One of the prime concerns in this chapter will be to demonstrate one way in which the analog output of a demodulator can be converted to digital form with the help of one or more decision thresholds. Another concern will be the consideration of the minimum ratio of signal energy to noise power density for which demodulation and synchronization are possible.

2.2 BINARY ANTIPODAL SIGNALING

As a simple illustration of the way in which analog information (plus noise) can be converted to digital form, consider a modulation scheme in which binary information is transmitted by means of antipodal waveforms; that is, if a 1 corresponds to the waveform $f(t)$, then a 0 corresponds to $[-f(t)]$. Assume that $f(t)$ has duration T and energy $\mathscr{E}$:

$$\int_0^T f^2(t)\, dt = \mathscr{E} \tag{2.1}$$

Furthermore, assume that we have stationary, additive, white Gaussian noise with sample function $n(t)$ of two-sided power density $N_0/2$ W/Hz, as shown in Figure 2.1. (The reader who is unfamiliar with the terminology used here will find the following texts very helpful: Chapters 3 and 4 of Wozencraft and Jacobs [2.1], Chapters 4 and 9 of Carlson [2.2];

Haykin [2.3]; Korn [2.4].) Finally, assume that the received waveform has optimum demodulation; i.e., the ratio of signal energy to noise power density is a maximum at the demodulator output. We achieve optimum demodulation by using a filter matched to the signal $f(t)$. That is, the filter has impulse response $h(t) = f(T-t)$. Thus, if we use this filter and sample its output at time $t = T$, then the (voltage) value designated by "Output" will have the largest possible ratio of signal energy to noise power density. In the illustration of the figure shown below, this ratio is $\mathscr{E} / N_0$, where $\mathscr{E}$ is the energy in $\pm f(t)$.

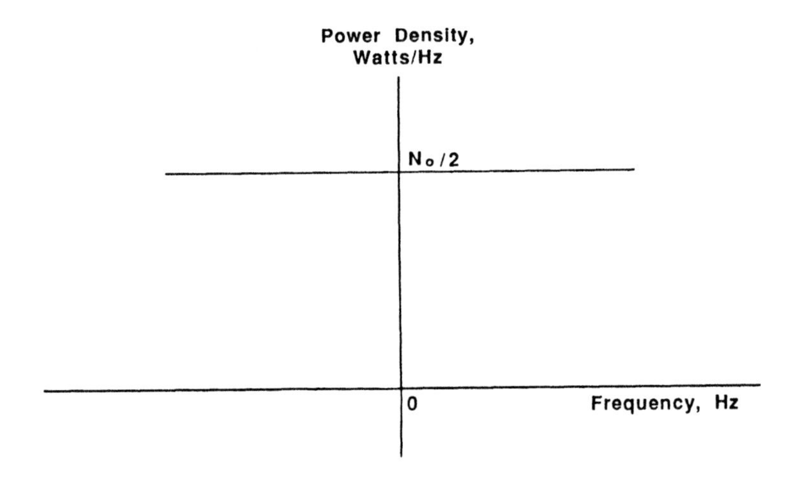

Figure 2.1 Two-sided white noise spectrum.

Let filter output waveform $\phi(t)$, consist of components $\phi_s(t)$ and $\phi_n(t)$, due to signal and noise, respectively:

$$\phi(t) = \phi_s(t) + \phi_n(t) \tag{2.2}$$

Because the matched filter is a linear system, superposition of inputs corresponds to superposition of outputs. Because the filter output is the convolution of the filter's impulse response with the input waveform, it follows that

$$[\pm f(t) + n(t)] * h(t) = \phi_s(t) + \phi_n(t) \tag{2.3}$$

or at time $t = T$,

$$\int_0^T [\pm f(T - \tau) + n(T - \tau)] f(T - \tau)\, d\tau = \phi_s(T) + \phi_n(T) \tag{2.4}$$

This expression consists of both deterministic and random components. That is, the matched filter output is a random variable. Due to linearity,

$$\pm \int_0^T f^2 (T - \tau)\, d\tau = \phi_s (T) \tag{2.5a}$$

$$\int_0^T n (T - \tau) f (T - \tau)\, d\tau = \phi_n (T) \tag{2.5b}$$

With the change of variable $\tau' = T - \tau$, (2.5a) becomes

$$\pm \int_0^T f^2 (\tau')\, d\tau' = \pm \mathscr{E} = \phi_s \tag{2.6}$$

That is, the signal component of the matched filter output is numerically equal to the signal energy. Although you obviously cannot evaluate (2.5b) explicitly, this is not a problem because all you really need is a statistical description of the filter output—in particular, the mean and variance and, if possible, the probability density of the output.

You can easily find the mean by using the symbol E to denote expected value:

$$\begin{aligned} E\,[\phi_s (T) + \phi_n (T)] &= E\left\{\int_0^T [\pm f^2 (T - \tau) \right. \\ &\qquad \left. + f (T - \tau) n(T - \tau)]\, d\tau\right\} \\ &= \pm \int_0^T E\,[f^2 (T - \tau)]\, d\tau \\ &\qquad + \int_0^T E\,[f(T - \tau) n(T - \tau)]\, d\tau \end{aligned} \tag{2.7}$$

The expected value of a deterministic function is the function itself. This fact, combined with (2.5a), shows that the first integral on the right-hand side of (2.7) has the value E_s. The second integral on the right can be written as follows:

$$\int_0^T f(T - \tau) E[n(T - \tau)]\, d\tau$$

Its value is zero because the noise has zero mean. Thus, the mean of the matched filter output is $\pm E_s$.

The calculation of the variance of the output is nearly as simple as that of the mean. By dropping the argument T, we obtain

$$\mathrm{var}(\phi_s + \phi_n) = E[(\phi_s + \phi_n - E_s)^2] \tag{2.8}$$

Collecting the deterministic portion of (2.8) gives

$$\mathrm{var}(\phi_s + \phi_n) = E[(\phi_s - E_s)^2] + E[\phi_n^2] + 2E[(\phi_s - E_s)\,\phi_n]$$

Because $\phi_s = E_s$ the first and third expected values are trivially zero. This leaves

$$E[\phi_n^2] = E[(\int_0^T n(T - \tau)f(T - \tau)\,d\tau)^2] \tag{2.9}$$

which may be written

$$\begin{aligned} E[\phi_n^2] &= E[\int_0^T ds \int_0^T n(T - \tau)n(T - s) \\ &\quad \cdot f(T - \tau)f(T - s)\,d\tau] \\ &= \int_0^T ds \int_0^T d\tau\, \{f(T - \tau)f(T - s) \\ &\quad \cdot E[n(T - \tau)\, n\,(T - s)]\} \end{aligned} \tag{2.10}$$

The expected value $E\,[n(T - \tau)\, n\,(T - s)]$ is simply the autocorrelation function $R_{nn}\,(\tau,s)$ of the noise. Assuming that the noise process is stationary, it can be shown that

$$E[n\,(T - \tau)\, n\,(T - s)] = (N_0/2)\,\delta\,(\tau - s) \tag{2.11}$$

for the noise power density of Figure 2.1, where $\delta\,(\tau - s)$ is the usual impulse or Dirac delta "function." After some manipulation, substitution of (2.11) into (2.10) yields

$$E[\phi_n^2] = E_s\, N_0/2$$

Therefore, the following is also true:

$$\mathrm{var}(\phi_s + \phi_n) = E_s\, N_0/2$$

Because the filter input is Gaussian, linearity (again!) tells us that the output must also be Gaussian, with a mean of $\pm\, E_s$, depending on the signal sent, and variance $E_s N_0/2$. The probability density function is thus one of the two curves shown in Figure 2.2.

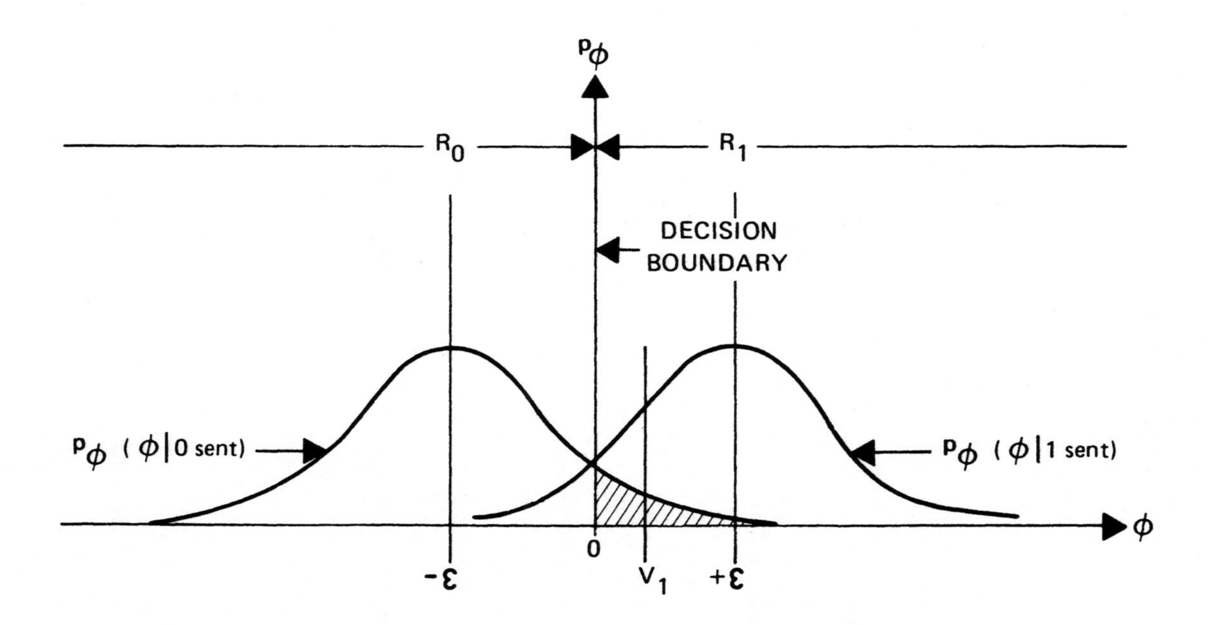

Figure 2.2 Probability density functions, decision regions, and error probability for binary signaling (0 and 1 are equally likely).

How do you use this information to decide whether a 0 or 1 was sent? For example, suppose you had matched filter output V_1 as shown in Figure 2.2. Intuitively, you would choose $+f(t)$ as the transmitted signal. Decision theory tells you that if 0 and 1 are equally probable at the transmitter, the optimum decision (in the sense of minimizing the probability of error) is obtained as follows:

1. Define decision regions R_0 and R_1 as the regions $\phi_s + \phi_n < 0$ and $\phi_s + \phi_n \geqslant 0$, respectively.
2. Choose 0 as the transmitted symbol if $\phi_s + \phi_n$ is in R_0. Choose 1 as the transmitted symbol if $\phi_s + \phi_n$ is in R_1.

Finally, what is the error probability associated with a demodulation decision? It is the probability of announcing that one symbol was sent when actually the other was sent. In terms of Figure 2.2, it is the probability of associating the filter output with the wrong density function, which, in turn, is equal to the area under the tail of the incorrect density function curve. Following the decision rule of the preceding paragraph, the received value V_1 would announce that a 1 had been sent. If, in fact, a 0 had been sent, the error probability would be the area of the shaded region of Figure 2.2. (The interested reader can find excellent treatments in Wozencraft and Jacobs [2.1] or Carlson [2.2] among others.)

This area is given by the analytical expression for the error probability given that 0 was transmitted:

$$P\,(e|0) = \int_0^{\infty} P\phi\,(\phi|0)\,d\phi$$

$$= \int_0^{\infty} \frac{1}{\sqrt{2\pi}\,\sqrt{\mathscr{E}N_0/2}}\exp\left[-\frac{(\phi+\mathscr{E})^2}{\mathscr{E}N_0}\right]d\phi \tag{2.12a}$$

Similarly,

$$P(e|1) = \int_{-\infty}^{0} \frac{1}{\sqrt{2\pi}\,\sqrt{\mathscr{E}N_0/2}}\exp\left[-\frac{(\phi-\mathscr{E})^2}{\mathscr{E}N_0}\right]d\phi \tag{2.12b}$$

Thus, the total error probability is

$$P(e) = P(0)\,P(e|0) + P(1)\,P(e|1)$$

A channel defined physically by symbol energy $\mathscr{E}$, noise power density N_0, and the decision process just described is usually represented mathematically by the *binary symmetric channel* (BSC) shown in Figure 2.3(a), with transition probability p equal to either of the integrals in (2.12a) and (2.12b). Note, however, that the input probabilities $P(0)$ and $P(1)$ need

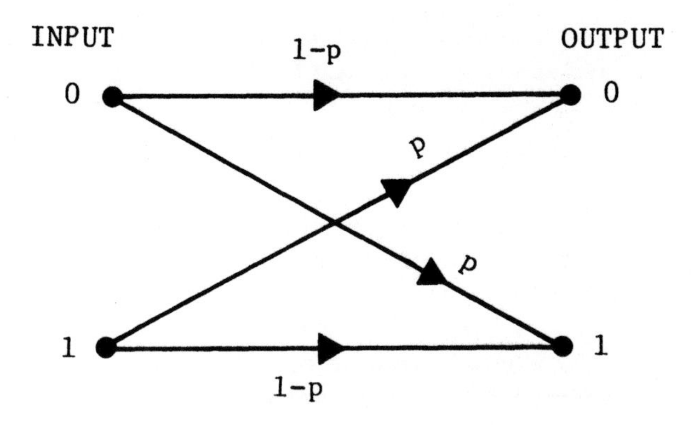

(a) Binary symmetric channel.

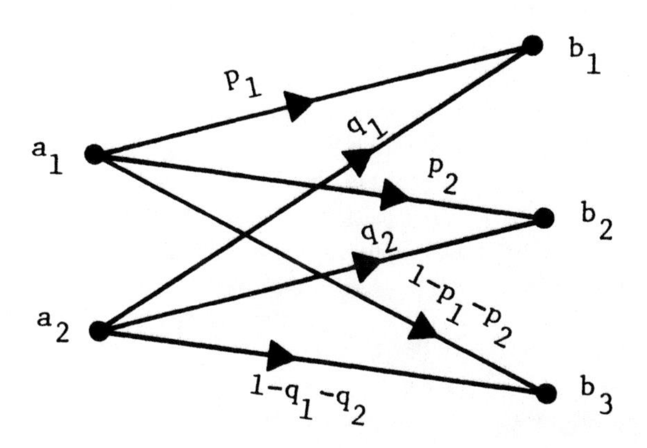

(b) Example of general discrete memoryless channel.

Figure 2.3 Models of discrete channels.

not be equal. Note also that transitions are independent from one digit to the next; that is, the channel is memoryless. Any change in these channel input probabilities will cause the optimum decision boundary (now at $\phi = 0$) to shift either right or left.

More generally, the number of inputs, number of outputs, and transition probabilities may take on any combination of values as long as the transition probabilities for each input sum to 1. This generalization of the BSC is called the *discrete memoryless channel* (DMC). A two-input, three-output channel with arbitrary transition probabilities is shown in Figure 2.3(b). A practical example of a DMC is a channel with binary inputs and, say, 3-bit (or 8-level) quantization at the output. Such a channel model is applicable to some convolutional decoding schemes, which have been implemented and are discussed later in this book.

2.3 BITS, BAUDS, SYMBOLS, AND CHIPS

The terms *bit, baud, symbol,* and *chip* are names of basic units used in digital communication. Because of their frequent use, it will be helpful to define and discuss them at this point. The usage here is intended to reflect current working definitions of these terms.

The word "bit" is usually used in one of two ways. The first, as in *data* or *information bit,* comes from the use as a unit of information. Thus,

$$I = -\log P_i \tag{2.13}$$

which is the amount of information conveyed by the occurrence of an event having probability P_i. For example, if 0s and 1s are equally likely to occur in a data stream, then $P_i = 1/2$. If the logarithm in (2.13) is taken to the base 2, then the units of I will be bits. A *data bit* is thus a 0 or 1 in a set of bits to which no redundancy has been added, giving $I = -\log_2 (1/2) = 1$ bit. This terminology, of course, has been applied for many years to a 0 or 1 being stored or processed by a digital computer. The second common use of the word "bit" primarily applies to digital communication, in which a 0 or 1 sent over a communication channel is usually called a *channel bit.* This is true whether or not the channel bit actually conveys one bit of information. The transformation from data bits to channel bits is usually referred to as *channel coding.* Because this operation could involve the introduction of redundancy, there will always be at least as many channel bits as there are data bits to represent a given message. The ratio of data bits to channel bits is called the code rate R.

For physical communication media, channel bits are represented by some sort of modulation, such as phase shift keying (PSK) or frequency shift keying (FSK). This modulation can be chosen to convey 1, 2, or even more channel bits per pulse. For example, choosing one of a set of two carrier frequencies results in binary FSK, which carries one channel bit per pulse. Each such pulse (FSK, PSK, *et cetera*)is referred to as a *channel symbol,* and we can talk about the channel symbol rate R_s of a communication system, where R_s is in channel symbols per second or *bauds.*

So far, then, we have talked about these quantities: code rate in data bits per channel bit; channel symbol or pulse rate in bauds; modulation rate in channel bits per channel symbol. The product of these three rates gives the actual information rate through the channel in data bits per second.

Example 2.1. Consider a system using 8-ary FSK modulation (i.e., choose one of eight possible carrier frequencies for each channel symbol) with 1200 such gated carrier pulses per second (i.e., 1200 baud). Assume that a rate $\frac{1}{2}$ error-control code is being used. There are $\log_2 8 = 3$ channel bits per channel symbol, $3 \times 1200 = 3600$ channel bits per second, and $\frac{1}{2} \times 3600 = 1800$ data bits per second. Also, there are $3 \times \frac{1}{2} = 1\frac{1}{2}$ data bits per channel symbol.

The concept of a *chip* is important insofar as it relates to the terms just discussed. Data bits, channel bits, and channel symbols are all concepts that arise in the context of digital communication taking place at data bandwidths (i.e., bandwidths whose values in hertz are of the same order of magnitude as the corresponding data rates in bits per second). The chip, on the other hand, is usually defined to be a basic pulse in a direct-sequence spread-spectrum communication system in which the bandwidth used between transmitter and receiver is very much greater than the data bandwidth (typically by a factor of 1000 or more). The quantity $10 \log_{10}$ (spread bandwidth/data bandwidth) is the spread-spectrum processing gain in decibels.

We often describe performance of modulation and coding schemes in terms of bit error probability P_B as a function of the ratio of bit energy E_B to white noise power density N_0. In virtually all instances, P_B and E_B refer to data or information bits. This provides a common basis for comparing coded and uncoded systems. Note that in the uncoded case, data bit and channel bit are identical. In some instances, energy per channel symbol E_s and channel-symbol error probability P_s will be specified. You must also take account of the number of data bits per channel symbol if you are using P_B and E_B as the basis for comparison. Although it is easy to convert between E_B and E_s, the relationship between P_B and P_s may not be as simple.

2.4 DEMODULATOR LIMIT ON CODING GAIN

Earlier in the chapter, we alluded to the potential difficulties in trying to use error-control coding in conjunction with an E_B/N_0 that is below some critical threshold value. We now address this problem more specifically. This question received considerable attention, especially in the context of a space channel or satellite relay channel in both analytical and simulation studies (Cahn [2.5], Heller [2.6], and Cahn, Huth, and Moore [2.7]).

The situation usually modeled, analyzed, or simulated builds on the discussion of binary antipodal signaling (i.e., biphase modulation) of a convolutionally encoded data stream. A typical system is shown in Figure 2.4, in which the more complete model of Figure 1.3 has been specialized to biphase modulation and convolutional coding, and the encryption and multiplexing have been eliminated in the interest of simplicity. The portions of the system of special interest are the three blocks pertaining to the insertion of reference bits, phase recovery, and ambiguity resolution. The need for phase recovery arises because this is a coherent system, which depends on knowledge of the carrier phase. Degradation of the received signal can result from oscillator drift, phase jitter, and relative motion of the transmitting and receiving terminals or relay satellite.

It is not our purpose here to go into the principles of Costas or phase-lock loop operation; these topics are discussed in Viterbi [2.8] and Gardner [2.9]. Rather, we shall simply summarize as typical the results given by Cahn, Huth, and Moore [2.7] for simulations of the system shown in Figure 2.4. This simulation is parameterized by the ratio of the loop bandwidth B_L to the channel symbol rate R_t. The bit error probability at the demodulator output is evaluated as a function of E_B/N_0. The standard of comparison is ideal coherent demodulation of the binary PSK signal plus noise. For $B_L = R_t/120$, a degradation from the ideal of about 0.4 dB is observed at an E_B/N_0 of about -1 dB, corresponding to bit error probability $P_B = 0.1$. When $B_L = R_t/12$, performance begins to degrade at about $E_B/N_0 = 5$ dB and deviate from the ideal by 0.5 dB at $E_B/N_0 \approx 3$ dB, $P_B \approx 0.03$. As B_L increases beyond $R_t/12$, degradation is rapid and severe.

There is obviously a limit on the usability of coding to compensate for errors due to low E_B/N_0. That is, for any modulation and coding combination, there exists a value or range of values of E_B/N_0 below which the demodulator output P_B is so degraded that the error-control scheme can hardly restore a tolerable (10^{-5} or better) output data bit error rate.

2.5 CONCLUSION

In summary, we observe that error-control coding represents a trade-off between a combination of cost, complexity, speed, and bandwidth and some combination of improved performance (P_B) and reduced transmitter power (E_B). The relative amount of each of these six system attributes given up or gained depends on the overall constraints present. For example, communication from a deep-space probe would probably be power-limited because of weight limitations on the space vehicle. This would call for a simple encoder and transmitter there and consequently a more complex, expensive, and probably slower (nonreal-time) decoder on earth to compensate for the very low signal-to-noise ratio characteristic of the deep space channel.

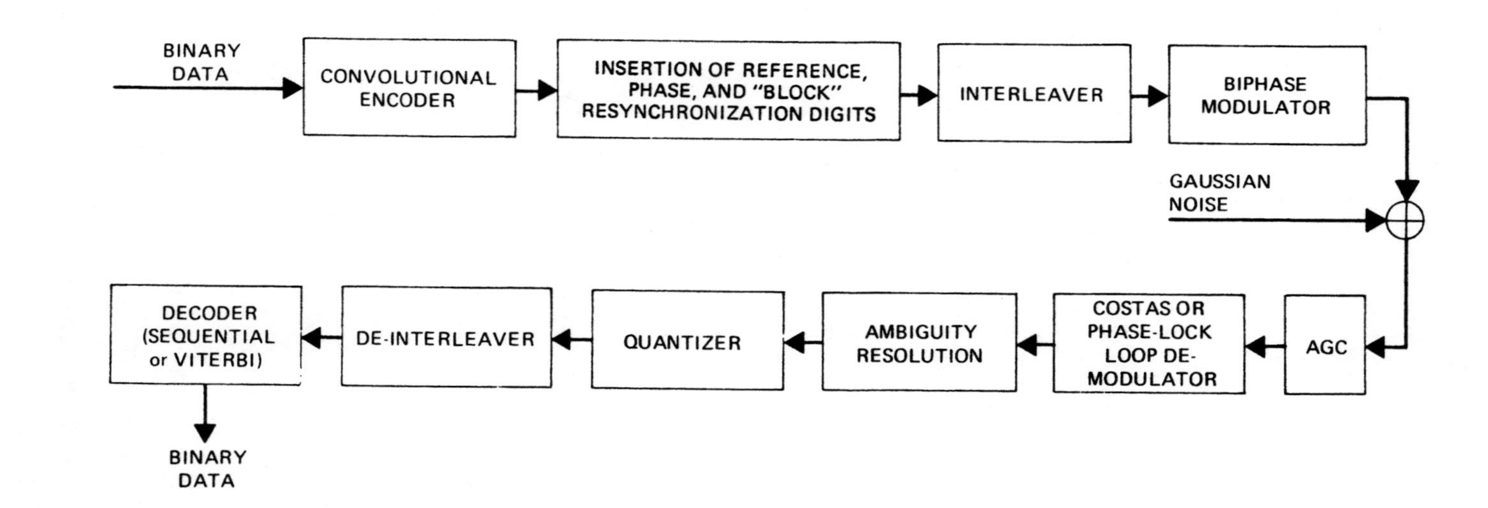

Figure 2.4 Model of a system featuring biphase modulation and convolutional coding.

Figure 2.5 shows the trade-off between error probability and signal-to-noise ratio for a block code and a number of constraint lengths of a convolutional code. Curves B, C, and D can each be compared to the reference curve A to show the gains in performance and savings in transmitter power made possible by error-control coding.

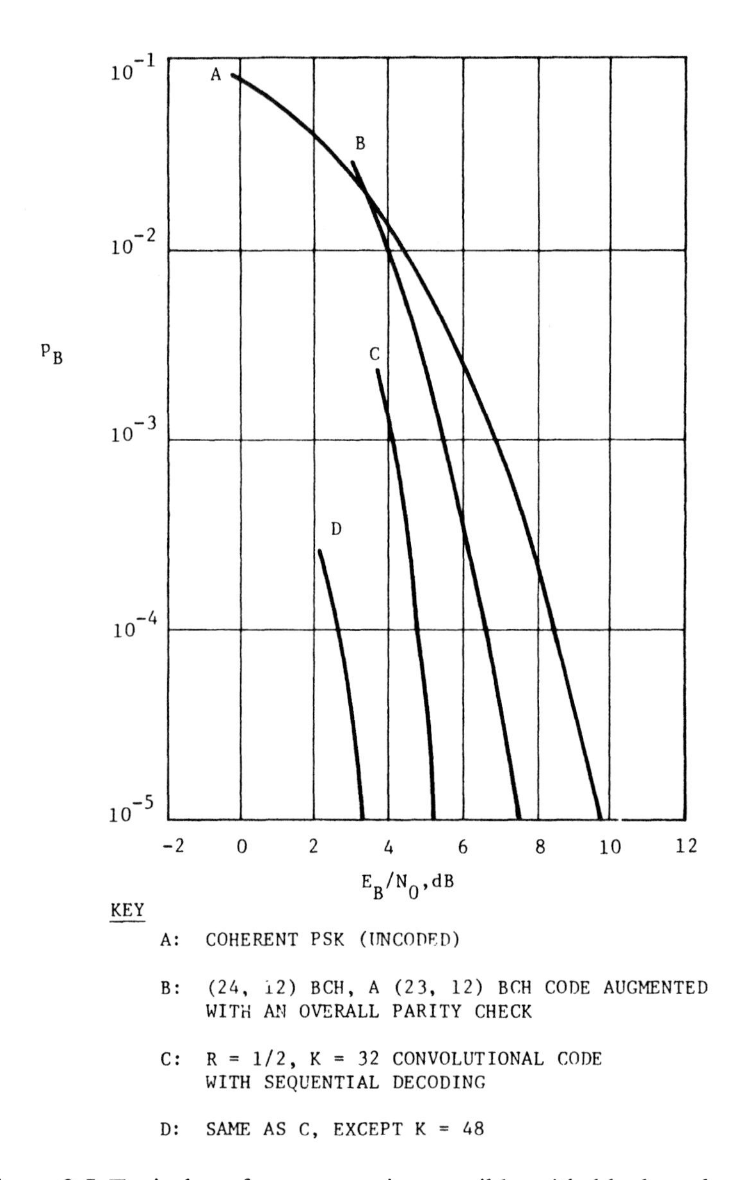

Figure 2.5 Typical performance gains possible with block and convolutional codes.

Chapter 3

Introduction to Parity-Check Codes

3.1 INTRODUCTION

We shall now study the construction and properties of error-detecting and error-correcting codes. In addition, we shall discuss how the use of vectors and matrices results in a compact representation of code words and their encoding and decoding operations. (Code words and operations on them can also be represented by polynomials, which are addressed in Chapter 4.) Many readers are probably familiar with vector and matrix operations from previous work in linear algebra or linear systems analysis (Appendix A). The representations of a code in terms of its generator matrix and parity-check matrix will be presented in the following sections. The interested reader may also refer to an approach used by Gallager [3.1] which, while using vectors and matrices, cleverly avoids defining and discussing vector spaces.

3.2 WHY STUDY ABSTRACT ALGEBRA?

In order to understand and use any but the simplest parity-check or repetition code, you need to develop a background in the concepts and operations of abstract algebra. While the words "abstract algebra" may imply to many engineers something totally useless, this is not the case. Abstract algebra enables us to identify the fundamental properties of such familiar mathematical systems as real numbers, rational numbers, and the set of all integers, as well as less familiar systems with a finite number of elements, such as the set {0, 1} with "binary arithmetic." Basic similarities and differences also become apparent. For example, you will see that {0, 1} and its arithmetic as just defined are mathematically more like the real numbers than they are like the integers. Examples of the kinds of systems to be discussed include groups, fields, and vector spaces. The real numbers, rational numbers, and the set $\{0, 1\}$ with binary arithmetic are examples of a field, while the integers using only addition are an example of a group. Because systems with a finite number of elements are used extensively in coding theory, it is a good idea to become comfortable with them as early as possible. For this reason, illustrative examples and exercises will emphasize groups, fields, and vector spaces having a finite number of elements.

3.3 PROPERTIES OF MATHEMATICAL SYSTEMS

Every mathematical system consists of one or more *elements* and at least one *operation* on any two of these elements. The same element may be used twice with respect to the operation. The elements might be a few or all of the integers, rationals, reals, or complex numbers, or they might be literals (*a, b, x, y, et cetera*). An operation could be ordinary addition or multiplication, or addition modulo some number m in which the result is the remainder left after dividing the ordinary arithmetic sum by m.

All mathematical systems must satisfy a small set of requirements. These requirements are as follows:

(G1) *Closure* under the specified system operations; i.e., if a and b are elements and $\times$ is an operation of system S, then $a \times b$ is also an element of S. (Note that $a \times b$ need not be equal to $b \times a$.)

(G2) Inclusion of an *identity* element i with respect to the operation $\times$, the effect of which is to leave any element unchanged. That is, $a \times i = i \times a = a$ for any a in S.

(G3) The existence for every element in S of an *inverse* a^{-1} (in S) with respect to the operation $\times$, such that $a \times a^{-1} = a^{-1} \times a = i$.

(G4) An *associative law*; that is, if a, b, and c are in S, then $(a \times b) \times c = a \times b \times c = a \times (b \times c)$ so that $a \times b \times c$ has an unambiguous meaning.

3.4 GROUPS

If the properties (G1) through (G4) are the only requirements on S, then S is called a *group*. If $a \times b = b \times a$ for all a and b in S, then the group is said to be *commutative* (or *abelian*) with respect to, or under, the operation $\times$.

Example 3.1. The set I of all positive and negative integers and 0, with the operation of ordinary addition, is a group. That is, the group operation

$\times$ is now the familiar operation "+". Clearly the sum of any two elements of I is again an element of I (e.g., $1 + 2 = 3$, which is in I; similarly, $2 + (-4) = -2$). The identity element is 0, since $0 + a = a + 0 = a$ for any number a in I. Finally, the inverse a^{-1} of any element a is $(-a)$, since $a + (-a) = 0$, the identity element. Note that this group is commutative because the order of the elements in any sum $a + b$ has no effect on the result.

Example 3.2. The simplest example of a group is the single element 0 under ordinary $+$: $0 + 0 = 0$, verifying closure, identity, and inverse; the associative law is trivially shown.

Example 3.3. Perhaps the next simplest example of a finite group consists of the elements 0 and 1 and the operation $\oplus$, which denotes addition modulo 2. That is, $1 \oplus 1 = 0$ because $1 + 1 = 2$, and $2 \div 2 = 1$ with remainder $= 0$. The complete addition table may be written

$\oplus$	0	1
0	0	1
1	1	0

where the entries in the first row and first column are the addends, and the entry of the intersection of any given row and column is the modulo-2 sum of the entries heading that row and column. This table clearly shows the closure property and the fact that 0 is the identity element under $\oplus$. What is the inverse of each element? That is, what value for a^{-1} satisfies $a \oplus a^{-1} = 0$ for $a = 0$ and for $a = 1$? A look at the table shows that each element is its own inverse: $0 \oplus 0 = 1 \oplus 1 = 0$.

3.5 FIELDS AND VECTOR SPACES

The group, defined and discussed in the preceding section, is interesting in its own right. It is also a necessary concept in defining two more complicated but more useful systems: *field* and *vector space.*

A *field* is a set of at least two elements and two operations, indicated here by $+$ and $\times$, that satisfies the following properties:

(F1) It is a commutative group under one operation (like $+$).

(F2) It is a commutative group under the other operation $\times$ except that 0 (the identity under $+$) has no inverse under $\times$.

(F3) A *distributive law* holds, which is defined by $a \times (b + c) = (a \times b) + (a \times c)$.

Note that property (F3) is a statement of linearity. The utility of this property will be illustrated in conjunction with Examples 3.8 and 3.9 below.

Example 3.4. The set of all positive and negative integers and 0, with the operations of ordinary addition + and ordinary multiplication ×, is not a field. Although this set is a commutative group under + (see Example 3.1), is closed under ×, and has 1 as the identity element under ×, only 1 and −1 have inverses under × (themselves). The distributive law given by property (F3) is satisfied for any set of integers or 0; for example, $2 \times (3 + (-1)) = 2 \times 2 = 4$, and $2 \times 3 + 2 \times (-1) = 4$.

Example 3.5. A very familiar example of a field is the set of rational numbers under ordinary addition and multiplication.

Example 3.6. The set {0, 1} is a field under ⊕ (as defined in Example 3.3) and ordinary multiplication. In this case, we have a *finite* (or Galois) field. Property (F1) has been verified in Example 3.2. All parts of property (F2) along with (F3) can be verified from the following multiplication table.

×	0	1
0	0	0
1	0	1

The notation GF(q) denotes the finite field of q elements.

Before defining a vector space, we shall first define a vector in the sense in which it is used in algebra and coding theory. We use the terminology "vector **v** having n components over a field F." This means simply that, for any integer $n \geqslant 1$,

$$\mathbf{v} = (v_1, v_2, v_3, \ldots, v_n) \tag{3.1}$$

is an n-tuple in which each v_i is an element of a field F. For example, $\mathbf{v} = (1, 0, 1)$ could be a vector over the real numbers, or it could be a vector over the finite field GF(2) having only the elements 0 and 1. Two operations are defined for vectors. For $\mathbf{v}_1 = (a_1, a_2, \ldots, a_n)$ and $\mathbf{v}_2$ $(b_1, b_2, \ldots, b_n)$, these operations are the following:

(V1) Addition of two vectors, defined by

$$\mathbf{v}_1 + \mathbf{v}_2 = (a_1 + b_1, a_2 + b_2, \ldots, a_n + b_n) \tag{3.2}$$

(V2) Multiplication by a scalar (that is, by an element of F), defined by

$$c\mathbf{v}_1 = (ca_1, ca_2, \ldots, ca_n) \tag{3.3}$$

where c and a_1 through a_n are elements of F, and multiplication of c by a_i is the $\times$ operation defined for F.

In particular, operation (V2) tells us that the negative of $\mathbf{v}_1$, $-\mathbf{v}_1$ is $(-1) \times \mathbf{v}_1$ and is given by

$$-\mathbf{v}_1 = (-a_1, -a_2, \ldots, -a_n)$$

Finally, the inner product (or dot product) of two vectors is defined as

$$\mathbf{v}_1 \cdot \mathbf{v}_2 = a_1 b_1 + a_2 b_2 + \ldots + a_n b_n \tag{3.4}$$

the result of which is again a scalar. In coding theory, this operation is used in calculating the vectors in the *null space* of a given vector space, i.e., a set of vectors $\mathbf{u}_i$ that satisfies $\mathbf{v} \cdot \mathbf{u}_i = 0$ for any $\mathbf{v}$ in the original space.

Example 3.7. Suppose we have $\mathbf{v}_1 = (2, 0, 1)$ and $\mathbf{v}_2 = (1, 1, 2)$ with vector components taken from the field F consisting of the elements 0, 1, and 2 with modulo-3 arithmetic. That is, addition and multiplication, respectively, are defined by

+	0	1	2
0	0	1	2
1	1	2	0
2	2	0	1

×	0	1	2
0	0	0	0
1	0	1	2
2	0	2	1

Therefore, the properties (V1) and (V2) for vectors are both illustrated by the following sum:

$$\begin{aligned} 2v_1 + v_2 &= 2 \times (2, 0, 1) + (1, 1, 2) \\ &= (2 \times 2, 2 \times 0, 2 \times 1) + (1, 1, 2) \\ &= (1, 0, 2) + (1, 1, 2) \\ &= (1 + 1, 0 + 1, 2 + 2) \\ &= (2, 1, 1) \end{aligned}$$

The inner product of $\mathbf{v}_1$ and $\mathbf{v}_2$ is

$$\begin{aligned} \mathbf{v}_1 \cdot \mathbf{v}_2 &= 2 \times 1 + 0 \times 1 + 1 \times 2 \\ &= 2 + 0 + 2 \\ &= 1 \end{aligned}$$

It is now possible to define a *vector space over a field F*. It is a commutative group under + (with components from F, and + as defined in F) for which addition of elements (vectors) is defined by property (V1). It is closed under the operation of scalar multiplication defined by property (V2). Furthermore, the two distributive laws

$$a(\mathbf{v}_1 + \mathbf{v}_2) = a\mathbf{v}_1 + a\mathbf{v}_2$$

$$(a + b)\mathbf{v}_1 = a\mathbf{v}_1 + b\mathbf{v}_1$$

and the associative law

$$a(b\mathbf{v}_1) = (ab)\mathbf{v}_1$$

all apply, where a and b are scalars. The additive identity is $\mathbf{0} = (0, 0, \ldots, 0)$, and the additive inverse $\mathbf{v}^-$ of vector $\mathbf{v}$ is

$$\mathbf{v}^- = (a_1^-, a_2^-, \ldots, a_n^-) \tag{3.5}$$

where a_i^- is the additive inverse of a_i in F.

The importance of the distributive laws and associative law is that they define and thus permit "moving a scalar (or vector) across the parenthesis" in situations involving the product of a vector and a scalar. These three laws will now be illustrated by examples. Let $\mathbf{v}_1 = (-1, 4)$, $\mathbf{v}_2 = (2, 3)$, $a = 3$, and $b = -2$.

Example 3.8

$$a(\mathbf{v}_1 + \mathbf{v}_2) = 3 \times [(-1, 4) + (2, 3)]$$

$$= 3 \times (1, 7) = (3, 21)$$

$$a\mathbf{v}_1 + a\mathbf{v}_2 = 3 \times (-1, 4) + 3 \times (2, 3)$$

$$= (-3, 12) + (6, 9) = (3, 21)$$

Example 3.9

$$(a + b)\mathbf{v}_1 = (3 - 2) \times (-1, 4) = 1 \times (-1, 4) = (-1, 4)$$

$$a\mathbf{v}_1 + b\mathbf{v}_1 = 3 \times (-1, 4) + (-2) \times (-1, 4)$$

$$= (-3, 12) + (2, -8) = (-1, 4)$$

Example 3.10

$$a(b\mathbf{v}_1) = 3 \times [(-2) \times (-1, 4)] = 3 \times (2, -8)$$
$$= (6, -24)$$
$$(ab)\mathbf{v}_1 = [3\,(-2)] \times (-1, 4) = (-6) \times (-1, 4)$$
$$= (6, -24)$$

The reader who says "That's obvious" should consider that these operations are "obvious" only for the familiar case in which $\mathbf{v}_1$ and $\mathbf{v}_2$ are replaced by scalars, giving ordinary field operations throughout. The extension of these notions to nonscalar quantities certainly seems natural enough but, as you may have learned by experience, intuition is helpful but not always sufficient or even correct. In the case of any new system such as the vector space here, all properties must be defined anew—in this case, in terms of the more familiar field over which the vector space is defined.

Example 3.11. The set of all triples (a_1, a_2, a_3) over the field F defined in Example 3.7 is a vector space V_1. Because each component can assume any of three values, there are $3^3 = 27$ distinct vectors in V_1.

Example 3.12. The set of those triples defined in Example 3.11 having one component (say a_3) held equal to 0 is also a vector space (call it V_1'), in this case with $3^2 = 9$ elements or vectors. These are (0, 0, 0), (0, 1, 0), (0, 2, 0), (1, 0, 0), (1, 1, 0), (1, 2, 0), (2, 0, 0), (2, 1, 0), (2, 2, 0).

The elements of the vector space V_1 of Example 3.11 can be constructed from those just listed for V_1' by letting the last component of each vector assume the values 1 and 2 as well as 0. We say that V_1' is a *proper subspace* of V_1. That is, V_1', as noted, is itself a vector space, and its vectors constitute only a fraction of all the vectors in V_1.

Listing all the elements of a vector space is an exhaustive (and possibly exhausting) method of describing the space. Isn't there a better way of doing this? The answer is a definite yes. A closer look at the nine vectors just listed above shows interesting properties.

Some vectors are simply multiples of others; for example, (0, 2, 0) = 2 × (0, 1, 0), and (1, 2, 0) = 2 × (2, 1, 0). (Remember that our vector space is over GF(3)). Thus, the vector on the right-hand side of each of these two equalities could be used to generate the one on the left side. The same scalar-multiple relationship holds for (1, 0, 0) and (2, 0, 0), and for (1, 1, 0) and (2, 2, 0). Thus, at least four vectors could be omitted from the original list of nine without causing any loss in the ability to represent the entire vector space. Furthermore, (0, 0, 0) is not needed in our compact list of vectors since it can be generated as the sum of any vector $\mathbf{v}_1$ and its additive inverse $\mathbf{v}_1^-$ (e.g., (1, 1, 0) and (2, 2, 0)). This leaves four out of the original nine vectors of V_1': (0, 1, 0), (1, 1, 0), (2, 1, 0), and (1, 0, 0). Of these, notice the following:

$$(2, 1, 0) = 2 \times (1, 0, 0) + (0, 1, 0)$$

$$(1, 1, 0) = (1, 0, 0) + (0, 1, 0)$$

That is, (2, 1, 0) and (1, 1, 0) can each be written as a sum of scalar multiples, or *linear combination,* of (1, 0, 0) and (0, 1, 0). Thus, all nine elements can be generated by operating on (1, 0, 0) and (0, 1, 0), neither of which can be produced from the other. These two vectors are an example of a pair of *linearly independent* vectors. In general, a set of vectors $\mathbf{v}_1$, $\mathbf{v}_2$, . . . , $\mathbf{v}_m$ is linearly independent if no vector, say $\mathbf{v}_i$, in the set is expressible as a linear combination of other vectors in the set.

$$\mathbf{v}_i = a_1\mathbf{v}_1 + a_2\mathbf{v}_2 + \ldots + a_{i-1}\mathbf{v}_{i-1} + a_{i+1}\mathbf{v}_{i+1} + \ldots + a_m\mathbf{v}_m$$

Furthermore, because the entire nine-element vector space can be generated by taking linear combinations of (1, 0, 0) and (0, 1, 0) and no others, these two vectors are said to form a *basis* for the subspace. More generally, a basis is a set of linearly independent vectors from which all vectors of a vector space can be generated as linear combinations. Thus, (1, 0, 0) and (1, 1, 0) are also a basis for V_1'. Figure 3.1 shows these two bases.

You have just seen by example and some labor that a vector space can be completely represented by a rather small subset of its elements. You have also seen a cumbersome way of determining such a subset. A compact way of writing down such a basis is as rows of a matrix:

$$A_1 = \begin{bmatrix} 1 & 0 & 0 \\ 0 & 1 & 0 \end{bmatrix} \quad \text{or} \quad A_2 = \begin{bmatrix} 1 & 0 & 0 \\ 1 & 1 & 0 \end{bmatrix}$$

3.6 LINEAR CODES

By definition, a *linear code* is a vector space, and each code word is a vector. Thus, matrix representation of a code is an ideal way to describe the code compactly and completely. A matrix whose rows are a basis for the vector space whose vectors make up the code is called a *generator matrix* for the code.

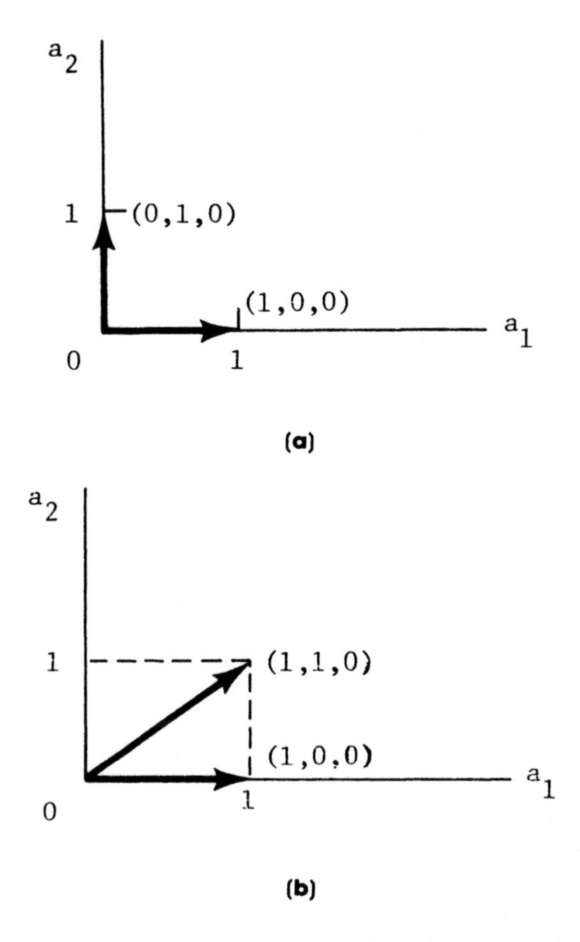

Figure 3.1 Two bases for the subspace V_1' of examples 3.11 and 3.12.

Because it is often useful to be able to determine a basis (i.e., to construct a generator matrix) given a set of vectors that constitutes a code, a systematic and efficient procedure is presented in Appendix A. Example A.5 illustrates the technique using the vector space of Example 3.12. The number of nonzero rows remaining after the matrix reduction described in this appendix is the number of linearly independent vectors in the corresponding vector space. This number is called the *dimension* of the vector space. If k is the dimension of a vector space and n is the number of components in a vector, then $k \leqslant n$. A matrix consisting only of these nonzero rows, in this case the matrix

$$G = \begin{bmatrix} 1 & 0 & 0 \\ 0 & 1 & 0 \end{bmatrix}$$

is a generator matrix for the code, the words of which constitute the vector space.

Example 3.13. As another example of a generator matrix, consider the following, this time with elements taken from the field consisting of the elements 0 and 1 with binary arithmetic as defined and discussed earlier in this chapter.

$$G_1 = \begin{bmatrix} 1 & 1 & 0 & 0 & 1 \\ 0 & 1 & 1 & 1 & 1 \\ 0 & 0 & 1 & 0 & 1 \end{bmatrix}$$

The vector space V_1 corresponding to G_1 has the dimension $k = 3$, with $n = 5$. How many code words are there in the code generated by G_1? That is, how many vectors are in the vector space for which the rows of G_1 are a basis? As obvious lower and upper bounds on this number, we have, respectively, $3 = k$ and $32 = 2^n$. To determine the actual number, recall that the vector space is generated by forming all linear combinations (with coefficients from the field over which the vectors are defined) of the basis vectors. For basis vectors $\mathbf{v}_1$, $\mathbf{v}_2$, and $\mathbf{v}_3$, we can write an arbitrary vector $\mathbf{v}$ of V_1 as

$$\mathbf{v} = a_1\mathbf{v}_1 + a_2\mathbf{v}_2 + a_3\mathbf{v}_3$$

Because each a_i can assume either of two values, there are $2^3 = 8$ possible combinations of a_1, a_2, and a_3; that is, eight code words are generated by G_1. In general, if there are k rows in a generator matrix, then there are 2^k and q^k code words when G_1 has elements from GF(2) and GF(q), respectively.

An interesting alternative way of looking at the generator matrix (and one which eliminates the need to discuss vector spaces) has been neatly presented by Gallager [3.1], whose method provides added insight into the significance of the generator matrix. First transform G_1 to

$$G_1' = \left[\begin{array}{ccc|cc} 1 & 0 & 0 & 1 & 1 \\ 0 & 1 & 0 & 1 & 0 \\ 0 & 0 & 1 & 0 & 1 \end{array}\right] = [I_3 | P] \tag{3.6}$$

by adding the third row of G_1 to the second row and then adding the newly calculated second row to the first row. Here I_3 (in general, I_k) is the 3 × 3 (or $k \times k$) identity matrix, and P is a $k \times (n - k)$ matrix, the properties of which will become apparent shortly. Now define a row vector **u** (which can be thought of as a *message*):

$$\mathbf{u} = (u_1, u_2, u_3)$$

and let the row vector **x** be defined by

$$\mathbf{x} = \mathbf{u}\, G_1'$$

Writing out this last equation in full gives

$$(x_1, x_2, x_3, x_4, x_5) = (u_1, u_2, u_3, u_1 + u_2, u_1 + u_3) \qquad (3.7)$$

Notice that the first $k(=3)$ components of **x** are identically equal to the k components of **u**, while the remaining $n - k$ $(= 5 - 3)$ components of **x** are linear combinations of components of **u**. Thus, the effect of G_1' on **u** has been to reproduce **u** and then append to it a number of additional digits. These additional digits are called *check digits*, or parity checks in the case of GF(2); they represent redundant information about the original vector **u**. In fact, it is useful here to write **x** as

$$\mathbf{x} = (u_1, u_2, u_3, c_1, c_2) \qquad (3.8)$$

In general, $(u_1, \ldots, u_k, c_1 \ldots, c_{n-k})$ calls attention to the nature of the various digits of **x** that we have just been discussing. The vector **x** is thus the code word corresponding to message **u**.

In fact, a code is *systematic* if each information digit appears as one of the digits of the code word **x**. This is the case for the matrix G_1'. Because every matrix can be reduced by row operations to a specific triangular form (as discussed in Section A.4), it follows that every linear code is equivalent to a systematic code. For any linear code (systematic or not), the information vector **u** uniquely determines the code word **x**. An example of a nonsystematic code is given by matrix G_1, for which $x_1 = u_1$, $x_2 = u_1, + u_2$, $x_3 = u_2 + u_3$, $x_4 = u_2$, $x_5 = u_1 + u_2 + u_3$, so that the code is now *nonsystematic*.

A code in which a k-component message **u** is represented by an n-component code word **x** is commonly referred to as an *(n, k) code*. In terms of the generator matrix, n is the number of columns and k is the number of rows. Finally, in terms of the corresponding vector space, n is the number of components in any vector and k is the dimension of the space.

We shall now develop one final important relationship between the check digits and the message digits. This relationship involves the well-known *parity-check matrix H*. Let g_{ij} be the element in the ith row and jth column of a generator matrix G. Equations (3.7) and (3.8) show that for the (5, 3) systematic code that we have been considering, the following relationships hold:

$$c_1 = x_4 = u_1 + u_2 \quad c_2 = x_5 = u_1 + u_3 \tag{3.9}$$

which may be rewritten, because of (3.7) and the properties of modulo-2 addition, as

$$x_1 + x_2 + x_4 = 0 \quad x_1 + x_3 + x_5 = 0 \tag{3.10}$$

Equations (3.10) have the following matrix form:

$$[x_1 x_2 x_3 x_4 x_5] \begin{bmatrix} 1 & 1 \\ 1 & 0 \\ 0 & 1 \\ \hline 1 & 0 \\ 0 & 1 \end{bmatrix} = [0 \quad 0] \tag{3.11}$$

from which it is clear that the top three rows of the 5×2 matrix are P as defined by (3.6), and the bottom two rows are simply I_2.

What code vectors x satisfy (3.11)? To answer this question, we first rewrite (3.10) in the form:

$$\begin{aligned} x_1 + x_2 \quad\quad + x_4 \quad\quad &= 0 \\ x_1 \quad\quad + x_3 \quad\quad + x_5 &= 0 \end{aligned} \tag{3.12}$$

This is a system of two equations in five unknowns, so the solution contains three arbitrary parameters. For the system of equations as written, it is convenient to let these parameters be simply x_1, x_2, and x_3, giving

$$\begin{aligned} x_4 &= x_1 + x_2 \\ x_5 &= x_1 + x_3 \end{aligned} \tag{3.13}$$

There are 2^3 possible sets of values x_1, x_2, x_3 where $\mathbf{u} = (x_1, x_2, x_3)$; thus there are also 2^3 code vectors, found most easily from (3.13). For example, if $x_1 = 1$, $x_2 = 1$, and $x_3 = 0$, then $x_4 = 1 + 1 = 0$ and $x_5 = 1 + 0 = 1$, so (1, 1, 0, 0, 1) is a code vector. All vectors found in this way will, of course, be generated by matrix G_1' of (3.6). The vector just found is, in fact, the sum of the first two rows of G_1'.

It is not difficult to generalize the example just concluded to an arbitrary field (bearing in mind our assumption of a systematic code, however, so that $x_i = u_i$ for $i = 1, 2, \ldots, k$). We have G given by

$$G = \begin{bmatrix} 1 & 0 \ldots 0 & g_{1,k+1} & g_{1,k+2} \cdots & g_{1,n} \\ 0 & 1 \ldots 0 & g_{2,k+1} & g_{2,k+2} \cdots & g_{2,n} \\ \cdot & \cdot \quad \cdot & \cdot & \cdot & \cdot \\ \cdot & \cdot \quad \cdot & \cdot & \cdot & \cdot \\ \cdot & \cdot \;\cdot\cdot & \cdot & \cdot & \cdot \\ 0 & 0 \ldots 1 & g_{k,k+1} & g_{k,k+2} \cdots & g_{k,n} \end{bmatrix} = [I_k P] \qquad (3.14)$$

Thus, the check-digit portions of the equation $\mathbf{x} = \mathbf{uG}$ become

$$\begin{aligned} c_1 &= x_{k+1} = g_{1,k+1}\, x_1 + g_{2,k+1}\, x_2 + \ldots + g_{k,k+1}\, x_k \\ c_2 &= x_{k+2} = g_{1,k+2}\, x_1 + g_{2,k+2}\, x_2 + \ldots + g_{k,k+2}\, x_k \\ &\ldots \\ c_{n-k} &= x_n = g_{1,n}\, x_1 + g_{2,n}\, x_2 + \ldots + g_{k,n}\, x_k \end{aligned} \qquad (3.15)$$

Taking a typical check digit $c_j = x_{k+j}$ and rewriting its equation as

$$\sum_{i=1}^{k} g_{i,k+j} x_i - x_{k+j} = 0 \quad j = 1, 2, \ldots, n - k \qquad (3.16)$$

The reader can verify that (3.16) is equivalent to the matrix equation:

$$(x_1\, x_2 \ldots x_n) \begin{bmatrix} g_{1,k+1} & \cdots\; g_{1,n} \\ g_{2,k+1} & \cdots\; g_{2,n} \\ \cdot & \\ \cdot & \\ \cdot & \\ g_{k,k+1} & \cdots\; g_{k,n} \\ -1 & \ldots 0 \\ \cdot & \\ \cdot & \\ \cdot & \\ 0 & \ldots -1 \end{bmatrix} = (0 \ldots 0) \qquad (3.17)$$

or

$$\mathbf{x}\begin{bmatrix} P \\ -I_{n-k} \end{bmatrix} = \mathbf{0} \tag{3.18}$$

where P is the same matrix as in (3.14) and $\mathbf{0}$ is a row vector with $n - k$ components. We usually write

$$H = \begin{bmatrix} P \\ -I_{n-k} \end{bmatrix}$$

and call H the *check matrix.* In the usual case of coefficients (or matrix elements) from GF(2), the minus signs in (3.16), (3.17), and (3.18) become plus signs and

$$H = \begin{bmatrix} P \\ I_{n-k} \end{bmatrix}$$

is the *parity-check matrix.*

3.7 THE CONCEPT OF DISTANCE

The need for creating codes in the first place stems from the desire to transmit and receive (or store and retrieve) data accurately in the presence of errors. Therefore, we must find ways of recovering the correct information digits in the presence of errors regardless of the information positions in the code word (or, in fact, even if the information digits do not appear explicitly, as in a nonsystematic code). The process of determining (actually, estimating) a set of transmitted information digits based only on received (and often corrupted) digits is called *decoding.*

The decoding operation is based on the fact that each unique sequence of information digits is encoded into a sequence of channel digits, which by design is different from the corresponding sequence of channel digits for any other sequence of information digits. A prime objective of code construction is to make these sequences of transmitted channel digits as different from one another as possible so that errors during transmission will not be sufficient to cause a received sequence to resemble the wrong transmitted sequence. This difference between any two channel digit sequences is called the *distance* between the two sequences. In the case of binary sequences, the number of bit positions in which two sequences of equal length differ is called their *Hamming distance.* Decoding can thus be based on finding the transmitted sequence that is at the smallest Hamming distance from the actual received sequence. We have other ways of describing the difference between a received and a transmitted sequence. One example is the *a posteriori* probability of a received sequence relative to a given transmitted sequence. Most examples in this book will use Hamming distance.

An important relationship between minimum distance and the capability of a code to correct or detect errors in the following: If the code has minimum distance $d_{\min}$ and can correct a maximum to t_C errors or detect a maximum of t_D errors, then

$$d_{\min} \geqslant 2t_C + 1 \tag{3.19}$$

and

$$d_{\min} \geqslant t_D + 1 \tag{3.20}$$

(Proof of these results is quite simple and is addressed in the Problem section). As far as notation is concerned, the symbol t is almost universally used to indicate error-correcting capability; and an (n, k), t-error-correcting code is sometimes referred to as an (n, k, t) code.

Thus, $v_1 = 10111$ and $v_2 = 01101$ have a Hamming distance of 3, written $d(v_1, v_2) = 3$. Another way of looking at Hamming distance is that it is the number of ones in the difference (or mod 2, position-by-position sum) of the two vectors—that is, the *weight* of the difference vector. In the example of this paragraph, $\mathbf{v}_1 - \mathbf{v}_2 = 11010$, which has weight 3. We write $w(\mathbf{v}_1 - \mathbf{v}_2) = 3$ and obtain the following relationship:

$$d(\mathbf{v}_i, \mathbf{v}_j) = w(\mathbf{v}_i - \mathbf{v}_j) \tag{3.21}$$

The minimum weight of any code word is the most commonly used measure of a code's effectiveness. Because a linear code is a vector space, the difference or sum of any two code words is also a code word. For this reason, the minimum distance between any two distinct code words can be determined by finding the minimum weight of any nonzero word in the code. The greater the Hamming distance between code words, the more difficult it is to change one into the other; thus, we look for codes with the largest possible minimum distance. It is not the purpose of this book to supply details as to how such codes are developed. However, distance properties of good codes, as well as the codes themselves in some cases, will be described and tabulated.

A useful way to think about the interrelationship between Hamming distance, error-correcting capability, and error-detecting capability is in terms of an n-dimensional space in which each code word is a point at distances d_1, d_2, *et cetera,* from other code words. Of course any n-tuple that is not a code word is also a point in this n-dimensional space. Consider the simple situation of a code word x_1, a neighboring code word x_2, and several n-tuples v_1, v_2, *et cetera,* which are not code words. If the code has minimum distance d_{min}, then any other code word is at least d_{min} away from code word x_1, and other vectors are at various distances from x_1.

These simple notions are illustrated in Figure 3.2, in which code words are shown as small circles, other vectors are shown as dots, and a small region of the n-dimensional space is projected into two dimensions. Because we wish to discuss a range of situations, especially the Golay code example, we present code words x_1 and x_2 a Hamming distance $d(x_1, x_2) = 8$ apart. The arcs shown in Figure 3.2 are the traces, in the plane of the page, of n-dimensional spheres centered on either x_1 or x_2 and enclosing all vectors at Hamming distance 1, 2, 3, . . . from one of the two code words.

Now consider the decoding rule discussed earlier in this section. Each received vector must be decoded into the code word nearest to it. For example, v_1 is closest to x_1 and would therefore be decoded into x_1. In general, if the code corrects all patterns of t or fewer errors, then all vectors within a "radius" t from a code word will be decoded into that word.

The occurrence of four errors in Figure 3.2 poses a problem: should a vector v_4 at a distance $d = 4$ from both x_1 and x_2 be decoded into x_1, x_2, or neither? In practice, a vector equidistant from all nearest code words would be decoded into none of them. Instead of this, a *detected* transmission error would usually be indicated. Thus, one way to use a code with $d_{min} = 8$ would be to correct all patterns of three or fewer errors and detect any pattern of four errors. Depending on the code, it may or may not be possible to correct any patterns of four or more errors.

Figure 3.2 shows $d_{min} = 8$. Clearly, transmission errors of one, two, and three channel bits will result in vectors that are inside or on spheres of radius one, two, and three, respectively, centered on the code word. The occurrence of five errors causes the vector to lie in a sphere centered on another code word, so that the vector may be decoded into this second code word for a decoding error. The relationship $d_{min} \geqslant 2t_C + 1$ is clearly seen in Figure 3.2.

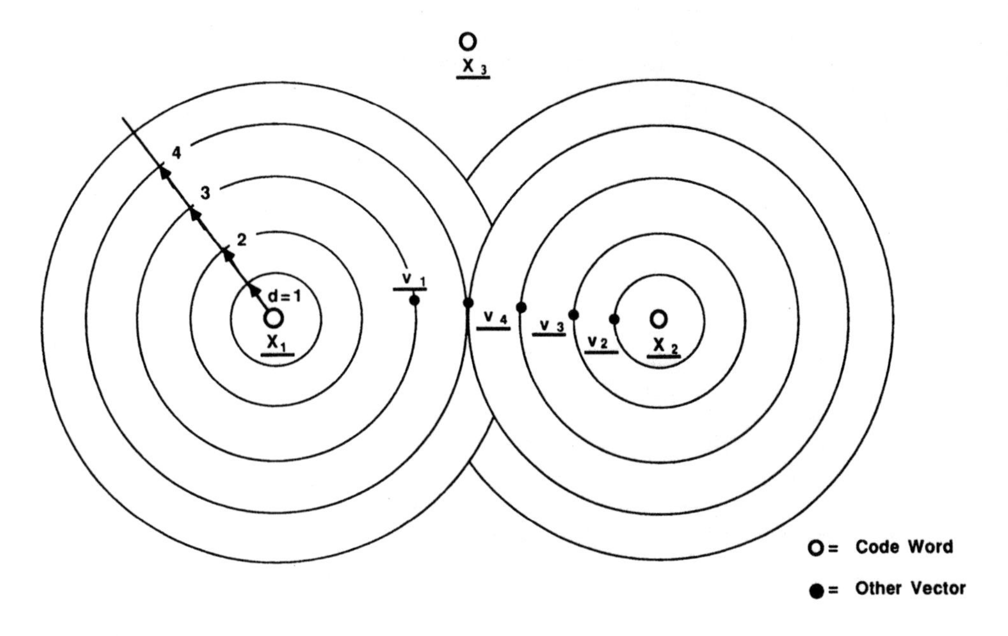

Figure 3.2 Combined detection and correction of errors.

Now suppose this same code is used only for error detection. Any vector that is not a code word is the result of one or more errors in transmission, and an undetectable error occurs only if one code word is changed by transmission errors into another code word. Thus, any number of bit errors less than $d_{\min}$ is always detectable. That is, $d_{\min} \geqslant t_D + 1$ detects t_D or fewer errors.

3.8 BASIC RESULTS IN WORD ERROR PROBABILITY

Suppose 0s and 1s are sent through the binary symmetric channel (BSC) derived in Chapter 2 and shown in Figure 2.3(a). This channel treats 0s and 1s exactly alike. In order to have any practical significance, transition probability p must obviously be less than 1/2. For $p < 1/2$, any symbol that passes through the channel is more likely to come out unchanged than it is to be transformed into the other value. A channel that acts on each symbol independently of all past and future symbols is called a *memoryless* channel. For any memoryless channel, the probability of a particular output sequence, given a particular code word transmitted, is equal to the probability of the corresponding sequence of errors introduced by the channel.

This sequence of errors is assumed to be independent of the word transmitted. That is, suppose that code word **x** is sent and sequence **y** is received. The difference between **x** and **y** is the error vector **e**, also made up of 0s and 1s and given by

$$\mathbf{e} = \mathbf{y} - \mathbf{x} \tag{3.22a}$$

which becomes

$$\mathbf{e} = \mathbf{y} + \mathbf{x} \tag{3.22b}$$

in GF(2).

Let us try to determine the most likely error sequences, in order of decreasing probability. Since each symbol is affected independently of all others, the probability of any error sequence is simply the product of the probabilities of the individual symbols which constitute it. Thus, if $\mathbf{e} = (e_1, e_2, \ldots, e_n)$, then

$$P(\mathbf{e}) = P(e_1)P(e_2) \ldots P(e_n) = \prod_{i=1}^{n} P(e_i) \tag{3.23}$$

where the notation Π means "product of." A 1 in an error vector means that the position in the transmitted word will be changed, giving an error there in the received vector.

Now apply equation (3.21) to several error vectors. For transition probability p, a symbol is received correctly with probability $1 - p$. Therefore, a transmitted vector consisting of n independent symbols will be received perfectly with probability $(1 - p)^n$. Any vector containing exactly one error has probability $p(1 - p)^{n-1}$. Similarly, a pair of errors will occur in a specific pair of positions in the vector with probability $p^2(1 - p)^{n-2}$. In general, the probability of a particular n-component vector (or n-symbol sequence) containing b errors is $p^b(1 - p)^{n-b}$. The important point here is that since $p < 1/2$, any sequence with no errors is most probable and, in particular, is more probable than any sequence with one error, which in turn is more probable than any two-error sequence, and so on. That is,

$$P(\text{no errors}) > P(1 \text{ error}) > P(2 \text{ errors}) > \ldots > P(n \text{ errors}). \tag{3.24}$$

Now consider the probability of all possible single-error sequences. Since a single error can occur in any of n positions, the probability of a sequence with exactly one error somewhere is $np(1 - p)^{n-1}$. It is easy to reason that two errors among n symbols can occur in exactly $n(n - 1)/2$ ways, so that the probability of two errors occurring in some way is

$$\frac{n(n-1)}{2}p^2(1-p)^{n-2}$$

In general, the total probability of all b-error sequences is

$$P(b \text{ errors}) = \binom{n}{b} p^b(1-p)^{n-b} \tag{3.25}$$

where the symbol $\binom{n}{b}$ denotes the number of ways in which n objects may be chosen b at a time without regard to their order. The value of $\binom{n}{b}$ is given by

$$\binom{n}{b} = \frac{n(n-1)(n-2)\ldots(n-b+1)}{b!} = \frac{n!}{b!\,(n-b)!}$$

Here, as usual, the factorial notation (!) is defined by

$$b! = b(b-1)(b-2)\ldots 3\cdot 2\cdot 1$$

3.9 DECODING PROCEDURES

This chapter on parity-check codes will conclude with a discussion of decoding procedures for discrete memoryless channels, with particular emphasis on the BSC. Consider first the probability of error P_e for a single channel symbol. The probability of a channel symbol error, conditional on symbol y being received, is $P(e|y)$. Without loss of generality, suppose $y = 0$ for a BSC. Then a decision error has been made on y if $x = 1$ was sent. Applying Bayes rule, we can write

$$\begin{aligned} P(e|0 \text{ received}) &= P(1 \text{ sent}|0 \text{ received}) \\ &= \frac{P(0 \text{ received}|1 \text{ sent})\, P(1 \text{ sent})}{P(0 \text{ received})} \end{aligned} \tag{3.26}$$

Now, $P(0 \text{ received}|1 \text{ sent})$ is just the channel transition probability p. Thus, setting $P(0 \text{ sent}) = P_0$ and $P(1 \text{ sent}) = P_1$, we have

$$\begin{aligned} P(0 \text{ received} &= P(0 \text{ received, } 0 \text{ sent}) + P(0 \text{ received, } 1 \text{ sent}) \\ &= P(0 \text{ received}|0 \text{ sent})\, P_0 + P(0 \text{ received}|1 \text{ sent})\, P_1 \\ &= (1-p)P_0 + pP_1 \end{aligned}$$

so that

$$P(e|0 \text{ received}) = \frac{pP_1}{(1 - p)P_0 + pP_1} \tag{3.27}$$

Note that if $P_0 = P_1$ (i.e., 0 and 1 equally likely to be sent), then equation (3.25) reduces to $P(e|0 \text{ received}) = p$.

With block codes, it is appropriate to consider the probability of decoding a block (received vector) incorrectly; that is, deciding on the wrong transmitted code word. The mathematical reasoning and formalism used in obtaining equation (3.25) are still valid, however, with symbols replaced by vectors in the derivation.

Now consider transmitted vectors (code words) $\mathbf{x} = (x_1, x_2, \ldots, x_n)$ and received vectors $\mathbf{y} = (y_1, y_2, \ldots, y_n)$, and suppose a particular sequence $\mathbf{y}_R$ is received. If vector $\mathbf{x}_k$ was actually sent, then the probability of correct decoding is

$$P_c = P(\mathbf{x}_k \text{ announced}|\mathbf{y}_R \text{ received})$$

A decision as to what was sent can be made on the basis of either of the following comparisons.

In the first comparison, compute $P(\mathbf{x}_i \text{ sent}|\mathbf{y}_R)$ for each code word $\mathbf{x}_i$ and announce that $\mathbf{x}_k$ was sent if

$$P(\mathbf{x}_k \text{ sent}|\mathbf{y}_R) > P(\mathbf{x}_i \text{ sent}|\mathbf{y}_R) \tag{3.28}$$

for all $i \neq k$. (A tie would have to be resolved separately.)

Clearly, this rule maximizes P_c and thus minimizes the probability of decoding error P_e, where

$$P_e = 1 - P_c$$

Thus, not surprisingly, this rule is called *minimum-error-probability decoding*.

How do you compute the probabilities used in equation (3.26)? In the interest of compactness and without loss of clarity, write $P(\mathbf{x}_i|\mathbf{y}_R)$ for $P(\mathbf{x}_i \text{ sent}|\mathbf{y}_R)$. The use of Bayes rule then gives

$$P(\mathbf{x}_i|\mathbf{y}_R) = \frac{P(\mathbf{y}_R|\mathbf{x}_i)P(\mathbf{x}_i)}{P(\mathbf{y}_R)} \tag{3.29}$$

so that for a given received vector $\mathbf{y}_R$, the decision rule is: announce that $\mathbf{x}_k$ was sent if, for all $i \neq k$,

$$P(\mathbf{y}_R|\mathbf{x}_k)P(\mathbf{x}_k) \geq P(\mathbf{y}_R|\mathbf{x}_i)P(\mathbf{x}_i) \tag{3.30}$$

Section 3.8 showed that the assumption of a memoryless channel leads to

$$P(\mathbf{y}_R|\mathbf{x}_i) = \prod_{j=1}^{n} P(y_{Rj}|x_{ij})$$

where y_{Rj} and x_{ij} are the j-th components of vectors $\mathbf{y}_R$ and $\mathbf{x}_i$, respectively. $P(\mathbf{x}_i)$ is just the probability of sending the code word $\mathbf{x}_i$. Thus, to implement (3.28), we announce that $\mathbf{x}_k$ was sent if

$$\prod_{j=1}^{n} P(y_{Rj}|x_{kj})P(\mathbf{x}_k) \geq \prod_{j=1}^{n} P(y_{Rj}|x_{ij})P(\mathbf{x}_i) \tag{3.31}$$

for all $i \neq k$.

Equation (3.31) shows that input message (or code word) probabilities must be known in order to carry out minimum-error probability decoding. When these probabilities are unknown, it is still possible to decode in a reasonable way if *maximum likelihood* (ML) decoding is used. The ML decoding rule is as follows. Given received vector $\mathbf{y}_R$, compute $P(\mathbf{y}_R|\mathbf{x}_i)$ for each code word $\mathbf{x}_i$. Then announce $\mathbf{x}_k$ sent if

$$P(\mathbf{y}_R|\mathbf{x}_k) \geq P(\mathbf{y}_R|\mathbf{x}_i) \tag{3.32}$$

for all $i \neq k$.

Comparison of (3.30) and (3.32) shows that the two decoding rules are equivalent if all code words $\mathbf{x}_i$ are equally probable, a condition which would have to be assumed in the absence of any other information and which could actually be the case.

The following three examples are intended to illustrate and contrast the decoding rules of (3.30) and (3.32) and to show the impact of input message probabilities on the resulting decisions. All three examples assume the following characteristics: a memoryless BSC with crossover probability $p = 0.1$. (which, as you may know, represents an extremely noisy channel); input code words $\mathbf{x}_1 = 000$ and $\mathbf{x}_2 = 111$; received sequence $\mathbf{y}_R = 001$.

Example 3.14. With the conditions given in the preceding paragraph, let $P(\mathbf{x}_1) = P(\mathbf{x}_2) = 1/2$. Then

$$P(\mathbf{y}_R|\mathbf{x}_1) = (0.9)^2(0.1) = 0.081$$

$$p(\mathbf{y}_R|\mathbf{x}_2) = (0.1)^2(0.9) = 0.009$$

and so by the ML decoding rule $\mathbf{x}_1$ is announced as having been sent if $\mathbf{y}_R$ is received. In view of the equal input (code word) probabilities, the minimum-error probability rule leads to the same decision.

Example 3.15. Now let $P(\mathbf{x}_1) = 0.1$ and $P(\mathbf{x}_2) = 0.9$. As in Example 3.14, $P(\mathbf{y}_R|\mathbf{x}_1) = 0.081$ and $P(\mathbf{y}_R|\mathbf{x}_2) = 0.009$, so that the ML decoding rule gives the same decision ($\mathbf{x}_1$ sent) as in Example 3.14. Equation (3.30) for the minimum-error probability rule becomes

$$P(\mathbf{y}_R|\mathbf{x}_1)P(\mathbf{x}_1) = (0.081)(0.1) = 0.0081$$

$$P(\mathbf{y}_R|\mathbf{x}_2)P(\mathbf{x}_2) = (0.009)(0.9) = 0.0081$$

giving a toss-up. In this case, either $\mathbf{x}_1$ or $\mathbf{x}_2$ could be announced as the transmitted code word.

Example 3.16. Finally, let $P(\mathbf{x}_1) = 0.01$ and $P(\mathbf{x}_2) = 0.99$. The ML decision is obviously still unchanged, but note what happens in the minimum-error probability case:

$$P(\mathbf{y}_R|\mathbf{x}_1)P(\mathbf{x}_1) = (0.081)(0.01) = 0.00081$$

$$P(\mathbf{y}_R|\mathbf{x}_2)P(\mathbf{x}_2) = (0.009)(0.99) = 0.00891$$

Therefore, announce that $\mathbf{x}_2$ was sent, even though this implies two channel errors out of three digits transmitted. The reason, of course, is that now $\mathbf{x}_1$ is much less likely than $\mathbf{x}_2$ to have been transmitted.

3.10 THE STANDARD ARRAY AND SYNDROME DECODING

The ML decoding procedure just described can be implemented by computing for the received vector $\mathbf{y}_R$ the 2^k values of $P(\mathbf{y}_R|\mathbf{x})$ and picking out the largest one. This effort grows exponentially with code length n for constant code rate k/n. On the other hand, values of $P(\mathbf{y}_R|\mathbf{x})$ could be computed beforehand and stored for speedy look-up. However, there are 2^n values of $\mathbf{y}_R$ for each of 2^k values of $\mathbf{x}_R$, giving 2^{n+k} entries $P(\mathbf{y}_R|\mathbf{x})$ in storage.

Some improvement in these astronomical storage requirements is possible with a decoding table. These storage requirements are very great indeed. For example, even the modest set of values $n = 24$, $k = 12$ gives $2^{36} = (512)^4 \approx 6.25 \times 10^{10}$ stored values.

On the other hand, computing values of $P(\mathbf{y}_R|\mathbf{x})$ represent $2^{12} = 4096$ computations, each consisting of an exclusive-OR followed by a computation of the form $p^d\ (1 - p)^{n-d}$, where d is the Hamming weight of $\mathbf{y}_R$

$-$ **x** and p is the value (or estimate) of the BSC transition probability. Even at the relatively slow rate of 10^6 computations per second, computing the 2^{12} values of $P(\mathbf{y}_R|\mathbf{x})$ would take only about 4 milliseconds.

Assume an (n, k) binary code, and suppose a table of binary n-tuples is constructed in the following way:

1. Write down the 2^k code words as column headings.
2. Choose an n-tuple of smallest possible weight that is not in the code.
3. Add this n-tuple to each code word, giving a second row below the row of code words.
4. Choose another n-tuple, again of smallest possible weight, which does not appear in the table constructed up to this point.
5. Add this n-tuple to each code word, generating another row in the table.
6. Repeat steps 4 and 5 until the entire set of 2^n n-tuples appears in the table, arranged into 2^k columns with a code word heading each column.

The table constructed in this way (called the *standard array*) can be used for ML decoding; that is, whenever an n-tuple is received, it is located in the table and decoded as the code word heading the column in which it was found. A simple example of this procedure is provided below.

Example 3.17. Suppose $n = 4$, $k = 2$, and the code words are 0000, 0101, 1011, and 1110. This is a systematic code in which the first two digits are information and the last two are checks formed as follows: the first check digit is a repetition of the first information digit, and the second check digit is the sum of the two information digits. Arrange the code words in a row (row 1 of Table 3.1). Pick a minimum-weight vector not in the code, say 1000, and add it to each code word (row 2). Pick another minimum-weight vector 0100 not appearing in either of the first two rows and add it to each code word (row 3). Note that row 3 also contains 0001. Thus, the only remaining vector of weight 1 is 0010, which is used to generate row 4. Because $2^4 = 16$, no other entries are possible in Table 3.1.

Table 3.1

0000	0101	1011	1110
1000	1101	0011	0110
0100	0001	1111	1010
0010	0111	1001	1100

Now use Table 3.1 to decode. Any received vector will be decoded into the code word that heads the column in which that particular vector appears. This procedure thus decodes into the nearest code word, which is a maximum-likelihood procedure. Therefore, any received vector containing two or more errors will certainly be decoded into the wrong code word, as will a vector produced by error vector 0001. If, for example, 0001 were the actual error vector and 0000 had been transmitted, the received vector would be decoded to 0101 as the transmitted vector, thus giving an error. On the other hand, if error vector 1000 actually occurred when 1011 was transmitted, then received vector 0011 would be correctly decoded. Suppose the error vector was 1100, and 1110 was sent. Then 0010 would be received, which would be decoded into 0000, giving a decoding error.

The construction of Table 3.1 shows that this decoding technique requires storage that grows exponentially with code length n. However, the decoding computation is trivial (a one-step table look-up.)

As the final result in this discussion of the decoding of block codes, we use the standard array combined with a trivial computation to achieve a significant reduction in storage requirements at the decoder with no appreciable decrease in speed. To do this, it is necessary to introduce the notion of the *syndrome* of a vector. For received vector $\mathbf{y}$ and transmitted code word $\mathbf{x}$, the syndrome is simply

$$s = \mathbf{y}H = (\mathbf{x} + \mathbf{e})H = eH \tag{3.33}$$

because $\mathbf{x}H = \mathbf{0}$. That is, the syndrome is actually the result of applying the parity-check equations to the error vector.

This operation in itself is not much help, but the following result makes the syndrome a very useful quantity (see Peterson and Weldon [3.2]). Two vectors, $\mathbf{y}_1$ and $\mathbf{y}_2$ are in the same row of the standard array if and only if their syndromes are equal. Thus, if a vector is received through a noisy channel and its syndrome is computed, the identity of the error vector is immediately known (assuming, of course, that the code's error-correcting capability has not been exceeded). At this point, simply add this noise vector (modulo 2) to the received vector to obtain the transmitted code word.

The key to the reduction of storage requirements is that instead of storing the entire row of the standard array, only the first element of that row (i.e., the error vector) along with its (unique) syndrome must be stored. Even the modest array of Example 3.17 is appreciably reduced in size. Whereas the original array consists of $2^4 = 16$ entries (12 vectors plus the four code words), the new storage consists of the code words plus three error vectors plus three syndromes, for a total of 10 entries.

Peterson and Weldon [3.4] cite the case of a (100, 80) code, for which the standard array has 2^{100} entries whereas the storage for syndrome decoding requires about $2^{n-k} = 2^{20}$ entries (still a very large number).

The concept of the syndrome will also be used in conjunction with threshold decoding of convolutional codes.

3.11 THE GOLAY CODE AND RELATED MATTERS

The binary Golay code has $n = 23$, $k = 12$, and corrects all patterns of three or fewer errors. However, it has no capability whatsoever to correct any patterns of more than three errors. Because of this property, it is one of a very small number of so-called *perfect* codes. These codes are "perfect" in the sense that a t-error-correcting perfect code has $d_{\min} = 2t + 1$ and can correct all patterns of 1, 2, . . . , t errors but not even one pattern of more than t errors. In terms of the standard array, there will be one row for the code itself, $\binom{n}{1} = n$ rows of one-error vectors, $\binom{n}{2}$ rows of two-error vectors, . . . , $\binom{n}{t}$ rows of t-error vectors, and no other rows. Thus, for the Golay code it is true that

$$\binom{23}{0} + \binom{23}{1} + \binom{23}{2} + \binom{23}{3} = 2^{23-12}$$

Any code can be *extended* by appending one or more check bits to the original code. Thus, an (n, k) code can be extended to give an (n', k) code, where $n' > n$. In this vein, the extended (24, 12) Golay code is constructed by appending an overall parity bit to the original (23, 12) code. The extended code has several interesting features. Its rate is exactly one half, which may be helpful in certain cases. Because it has $d_{\min} = 8$, it retains the capability of the original (23, 12) code to correct all patterns of three or fewer errors, but it now is also capable of detecting all four-error patterns. The (24, 12) code will be used as the basis of the following discussion of combined error detection and correction. (A complete discussion of both the (23, 12) and (24, 12) codes is given in the book by MacWilliams and Sloane [3.3]. In addition to discussing the codes, the book by Michelson and Levesque [3.4] contains a detailed discussion of decoding Golay codes.)

Thus far, the discussion concerning the decoding of block codes has focused entirely on the usual processes of pure correction or pure detection. An interesting and useful extension of these two processes consists of using error correction and error detection in various combinations. For example, suppose you decide to correct only one error with the triple-error correcting code of Figure 3.2. How many additional errors can be detected? The answer is as follows: any number corresponding to a point lying between the two circles (in general, n-dimensional spheres) of radius = 1 that are centered on code words $\mathbf{x}_1$ and $\mathbf{x}_2$. Counting outward from either of these words shows that vectors at distances 2, 3, 4, 5 or 6 contain error patterns guaranteed to be detectable. The vector $\mathbf{v}_2$ at $d = 7$ from $\mathbf{x}_1$ is also at $d = 1$ from $\mathbf{x}_2$, causing it to be decoded into $\mathbf{x}_2$; such a vector therefore does not generally contain a detectable error pattern. In this situation, then, it is possible to correct single errors and detect all patterns from two to six errors.

Another possibility is to correct all patterns of one and two errors. Where does this lead, as far as detecting additional errors is concerned? Clearly, the vector $\mathbf{v}_3$ at $d = 5$ from $\mathbf{x}_1$ is at distance 2 from $\mathbf{x}_2$ and would therefore now be decoded into $\mathbf{x}_2$. Therefore, only patterns of three or four errors are guaranteed to be detectable, so that it is possible simultaneously to correct up to two errors and detect up to four errors. Note that if $d(\mathbf{x}_1, \mathbf{x}_2)$ were greater than d_{min}, it might be possible to correct or detect more than the minimum numbers discussed here, but this capability is not guaranteed. (This exploitation of the range of combinations of error correction and detection was recently suggested by Dr. Kenneth MacDavid of Harris Corporation, Rochester, NY, for incorporation into proposed FED-STD-1045 in conjunction with the (24, 12), $d = 8$, extended Golay code.)

3.12 ERRORS *VERSUS* ERASURES

Up to this point, the only type of corruption of transmitted data that we have discussed is the *error*. An error actually represents two unknown quantities in general: location and symbol value. In the binary case this simplifies to location alone. An *erasure*, on the other hand, has a known location but a completely unknown value. An erasure could arise, for example, if the demodulator output failed to pass some confidence level, giving rise to considerable uncertainty as to the value of the symbol or binary digit. The position in the code word would then be flagged at the decoder input as an erasure.

Chapter 4
Abstract Algebra and Cyclic Codes

4.1 INTRODUCTION

This chapter presents the minimum amount of the theory of finite fields and cyclic codes needed for an understanding of the properties of cyclic codes in general and (for work in later chapters) of BCH and Reed-Solomon codes in particular. The work of the present chapter builds on the material on finite fields introduced in Chapter 3. Results are stated without proof, as before, but are motivated and illustrated to an extent that should render them useful and meaningful to the reader.

4.2 PROPERTIES OF POLYNOMIALS OVER A FIELD

The Euclid division algorithm stated in Theorem 4.1 is important because of what it tells about the remainder resulting when one polynomial is divided by another. The theorem holds regardless of the degrees of the dividend and divisor polynomials involved. The notation deg $[f(x)]$ denotes the degree of the polynomial $f(x)$.

Theorem 4.1

Let $a(x)$ and $b(x)$ be polynomials of at least the first degree with coefficients in some field $\boldsymbol{F}$. Then unique polynomials $q(x)$ and $r(x)$ exist, also with coefficients in $\boldsymbol{F}$, such that

$$a(x) = q(x)b(x) + r(x) \tag{4.1}$$

where $q(x)$ may be of any degree but the degree of $r(x)$ is less than the degree of $b(x)$.

Note that the more familiar "division problem" form of (4.1) is

$$a(x)/b(x) = q(x) + r(x)/b(x)$$

where $\deg[q(x)] + \deg[b(x)] = \deg[a(x)]$ unless $\deg[b(x)] > \deg[a(x)]$, in which case $q(x) \equiv 0$. Of particular interest here will be the case $\boldsymbol{F} = \mathrm{GF}(p)$, where p is a prime number. Nonetheless, as an aid to initial understanding, the first example will use $\boldsymbol{F}$ = the rationals.

Example 4.1. Let $a(x) = x^3 + 1$, $b(x) = x^2 + 1$, and $\boldsymbol{F}$ = rational numbers. Then, by the following procedure (or by long division of polynominals), (4.1) becomes

$$x^3 + 1 = x(x^2 + 1) - x + 1 = x \cdot b(x) + (-x + 1)$$

so that $q(x) = x$ and $r(x) = -x + 1$.

Example 4.2. Let $a(x) = x^3 + 1$ and $b(x) = x^2 + 1$, but let $\boldsymbol{F} = \mathrm{GF}(2)$. Then

$$x^3 + 1 = x(x^2 + 1) + x + 1$$

so that $q(x) = x$ and $r(x) = x + 1$.

Example 4.3. Let $a(x) = x^2 + 1, b(x) = x^3 + 1$, and $\boldsymbol{F} = \mathrm{GF}(2)$. Because $\deg[b(x)] > \deg[a(x)]$, $q(x)$ must be 0, which forces $r(x) = x^2 + 1$.

Chapter 3 presented arithmetic modulo a prime number. In a similar way, it is possible to define operations modulo expressed as a polynomial with coefficients from some field $\boldsymbol{F}$. Consider polynomials $a(x)$ and $P(x)$. Thus, $a(x)$ modulo $P(x)$ simply means the remainder $r(x)$ that results when $a(x)$ is divided by $P(x)$. This relationship is written as follows:

$$a(x) = r(x) \bmod P(x) \tag{4.2}$$

which means that

$$a(x) = q(x)P(x) + r(x).$$

where $r(x)$ has degree less than the degree of $P(x)$.

Example 4.4. For the conditions of Example 4.2, there results

$$x^3 + 1 = x + 1 \quad \mathrm{mod}(x^2 + 1)$$

If $P(x) = a(x)$, then clearly $r(x) = 0$; thus for example,

$$x^3 + 1 = 0 \mod(x^3 + 1)$$

A polynomial $a(x)$ is defined to be *reducible* over a field $\boldsymbol{F}$ if there exist two polynomials $b(x)$ and $c(x)$, both of degree one or greater with coefficients in $\boldsymbol{F}$, such that

$$a(x) = b(x)c(x)$$

Note that another term for "reducible" might be "factorable." This definition excludes simply factoring out a field element to give $a(x) = ka_1(x)$, where k is an element of $\boldsymbol{F}$. A polynomial is called *irreducible* if it is not reducible. Irreducible polynomials are very important in the study of cyclic codes.

Example 4.5. Consider $a(x) = x^2 + 1$. If $\boldsymbol{F}$ is the rational or real numbers, $a(x)$ is certainly irreducible over $\boldsymbol{F}$. If $\boldsymbol{F}$ is the field of complex numbers, however, the result is

$$x^2 + 1 = (x + \mathrm{j}1)(x - \mathrm{j}1)$$

where $\mathrm{j} = \sqrt{-1}$. Finally, if $\boldsymbol{F} = \mathrm{GF}(2)$, $x^2 + 1 = (x + 1)^2$.

Another useful concept is the *monic* polynomial, in which the highest-degree term has a coefficient of unity. Also, an element α is a root of $a(x)$ with coefficients in $\boldsymbol{F}$ if $a(\alpha) = 0$. Note that α may not be an element of $\boldsymbol{F}$. (More precisely, α is a root of $a(x) = 0$, or α is a zero of $a(x)$.) Thus, in Example 4.5, $\alpha = \pm\, \mathrm{j}1$ is not in the reals, so that $x^2 + 1$ has no real roots. A larger field $\boldsymbol{F}_1$ which contains a field $\boldsymbol{F}$ is called an *extension field* of $\boldsymbol{F}$. Thus, the reals are an extension field of the rationals, and the complex numbers are an extension field of the reals.

The following theorem is often taken as an axiom in introductory algebra courses. It can, however, be derived from more fundamental principles and therefore is actually a theorem.

Theorem 4.2

An element α is a root of $a(x)$ if and only if $(x - \alpha)$ is a factor of $a(x)$. There are no more than n distinct roots of a polynomial $a(x)$ of degree n. (As discussed in the preceding paragraph, α may or may not be an element of the coefficient field of $a(x)$.)

4.3 SOME PROPERTIES OF GALOIS FIELDS

A small additional amount of algebraic theory is needed in order to discuss cyclic codes. This theory has to do specifically with finite, or Galois, fields.

First consider a set of polynomials of degree $n - 1$ or less, with coefficients from a *finite* field GF(p), where p is prime. Bearing in mind that operations on the coefficients take place in GF(p), let the operations on these polynomials be the following:

Operation PA: Polynomial addition; i.e., term-by-term.
Operation PMR: Multiplication of polynomials followed by reduction modulo expressed as some nth degree polynomial $P(x)$, that is irreducible over GF(p).

The result of applying the foregoing definitions of polynomials and operations is this key result: the set of polynomials generated in this way constitutes a field of p^n elements designated GF(p^n), which has the set of all possible polynomials resulting from operation PMR as just one of at least two representations.

Example 4.6. Suppose $P(x) = x^2 + x + 1$, which is irreducible over GF(2). Consider the following two polynomials over GF(2): $x + 1$ and $x^2 + 1$. Then, taking all polynomials mod $(x^2 + x + 1)$, there results

$$(x^2 + 1) + (x + 1) = (x^2 + x + 1) + 1 = 1$$

$$(x + 1)(x^2 + 1) = x^3 + x^2 + x + 1 = x(x^2 + x + 1) + 1 = 1$$

In fact, $x^2 + 1 = (x^2 + x + 1) + x = x$.

Note in this example that an alternative approach to the first sum yields $x^2 + x + (1 + 1) = x^2 + x$. Is this at odds with the result obtained the first time? No, because all results must be of degree less than the degree of $P(x)$. Thus, reduction of $x^2 + x$ modulo $(x^2 + x + 1)$ yields, as above,

$$x^2 + x = x^2 + x + 1 + 1 = 1 \tag{4.3}$$

In this case $p = 2$, $n = 2$, and the elements of GF(2^2) can be represented as 0, 1, α, and $\alpha + 1$, where α satisfies $P(\alpha) = 0$ so that $\alpha^2 = \alpha + 1$. This last relationship is the key to evaluating any expression involving the second and higher powers of α.

Now define the *multiplicative order* of an element α of $GF(p^n)$ to be the smallest integer μ (of the set of natural numbers, not from $GF(p^n)$) such that

$$\alpha^{\mu} = 1 \quad \text{in } GF(p^n) \tag{4.4}$$

If $\mu = p^n - 1$, then α is called a *primitive element* of $GF(p^n)$. In this case, it can be shown that the successive powers of α generate all nonzero elements of $GF(p^n)$. In addition, it can be shown that every Galois field contains at least one primitive element.

Example 4.7. Consider the field $GF(2^2)$, which has just been defined and discussed (note that now $p^n = 4$). The element α defined by $P(\alpha) = \alpha^2 + \alpha + 1 = 0$ is a primitive element of $GF(2^2)$. To verify this claim, simply compute $\alpha^2 = \alpha + 1$ and $\alpha^3 = \alpha(\alpha + 1)(\alpha + 1) + \alpha = 1$, thus showing that $\mu = 3 = p^n - 1$. Note that $\alpha + 1$ is also a primitive element: $(\alpha + 1)^2 = \alpha^2 + 1 = (\alpha + 1) + 1 = \alpha$, $(\alpha + 1)^3 = (\alpha + 1)\alpha = 1$.

Another important concept is that of a *minimal polynomial.* For our purposes, the following definition will suffice. The minimal polynomial $m_\alpha(x)$ over GF of α, which is an element of $GF(p^n)$, is the monic polynomial of lowest degree such that $m_\alpha(\alpha) = 0$.

Example 4.8. Denote the minimal polynomials of the elements 0, 1, α, $\alpha + 1$ of $GF(2^2)$ by $m_0(x)$, $m_1(x)$, $m_\alpha(x)$, and $m_{\alpha+1}(x)$, respectively. It is trivial to see that $m_0(x) = x$ and $m_1(x) = x + 1$. Since α was defined to be a zero of the irreducible polynomial $P(x)$, then $m_\alpha(x)$ must be of the degree ≥ 2. The polynomials x^2, $x^2 + 1$, and $x^2 + x$ are ruled out because each of these polynomials factors into a product of linear factors over GF(2). This leaves $P(x)$ as the only remaining candidate of degree 2, and obviously $P(\alpha) = 0$, so $m_\alpha(x) = x^2 + x + 1$. Finally, you can test the element $\alpha + 1$ to verify that x, $x + 1$, x^2, $x^2 + 1$, and $x^2 + x$ all fail as minimal polynomials for $\alpha + 1$, for the same reasons given for their failure for α. Finally, try $x^2 + x + 1$: $(\alpha + 1)^2 + (\alpha + 1) + 1 = (\alpha^2 + 1) + \alpha = 0$. Thus, $m_{\alpha+1}(x) = P(x)$, also.

The preceding example was easy to do using exhaustive procedures. If the degree of $m(x)$ is even slightly larger than 2, it becomes quicker to use the following result: let β be an element of $GF(2^m)$ and let $r > 0$ be the smallest integer such that $\beta^{e(r)} = \beta$, where $e(r) = 2^r$. Then $m_\beta(x) = \prod_{i=0}^{r-1} (x + \beta^{e(i)})$ is the minimal polynomial of β.

Example 4.9. The result just stated will be used to find the minimal polynomial of α^2, where $\alpha^2 = \alpha + 1$ in GF(2^2). We are looking for the smallest positive integer r for which $\beta^{2^r} = \beta$. Because $\beta = \alpha + 1$, then

$$\beta^2 = \beta^2 \alpha^2 + 1 = \alpha = \beta + 1$$

$$\beta^4 = (\beta^2)^2 = \alpha^2 = \beta$$

Therefore $r = 2$, and

$$m_{\alpha^2}(x) = (x + \beta)(x + \beta^2)$$

$$= (x + \alpha^2)(x + \alpha) = (x + \alpha + 1)(x + \alpha)$$

$$= x^2 + x + \alpha(\alpha + 1) = x^2 + x + 1$$

as expected.

Because of its usefulness in the construction of cyclic codes, one last property of a Galois field will be stated. Let $f(x)$ be an irreducible polynomial of degree m over GF(p), and let α be a root of $f(x) = 0$. Then α is in GF(p^m), and the other roots of $f(x) = 0$ are $\alpha^p, \alpha^{p^2}, \ldots, \alpha^{p^{m-1}}$. For example, if $p = 2$ and $f(x) = x^4 + x + 1$, then $m = 4$, a root α of $f(x) = 0$ is an element of GF(2^4), and the other roots of $f(x) = 0$ are α^2, α^4, and α^8.

To tie together the ways in which a Galois field can be represented, and to see an example that is definitely nontrivial, consider the field GF(2^3) as determined by $P(x) = x^3 + x^2 + 1$ over GF(2). This polynomial is primitive; i.e., it is an irreducible polynomial of maximum degree that divides $x^7 - 1$, where $7 = 2^m - 1_e$. Let α be a root of $P(x) = 0$. Then $\alpha^3 = \alpha^2 + 1$ is used as the relationship to reduce powers of α greater than α^2. Table 4.1 gives three equivalent representations of GF(2^3)—as powers of α, as polynomials in x, and as vectors. Note that the degree of the terms increases from left to right. Addition of elements is most easily done using either the polynomial or the vector representation (term by term, or component by component, respectively). Bear in mind that all operations on coefficients are in GF(2).

Multiplication must be done with reference to the primitive element powers. For example, (001) $\times$ (101) corresponds to $\alpha^2 \times \alpha^3 = \alpha^5$, which corresponds to vector (110); (011) $\times$ (111) is $\alpha^6 \times \alpha^4 = \alpha^{10} = \alpha^3$, which is, once again, (101). Be particularly sure to notice that no simple rule applies for directly multiplying two vectors, nor does any rule apply for directly adding two powers of α.

Table 4.1 Representations of GF(2^3) $P(x) = 1 + x^2 + x^3$

as powers of a primitive element	*as polynomials modulo P(x)*	*as vectors*
$\alpha^{-\infty}$	0	000
α^0	1	100
α^1	x	010
α^2	x^2	001
α^3	$1 + x^2$	101
α^4	$1 + x + x^2$	111
α^5	$1 + x$	110
α^6	$x + x^2$	011

4.4 INTRODUCTION TO CYCLIC CODES

Cyclic codes are linear block codes (the subject of the first five chapters of this book). The special, defining property of cyclic codes is that each cyclic (end-around) shift of a code word is also a code word. Like any other (n, k) linear block code, a cyclic code can be represented by a generator matrix or a parity-check matrix, with each code word written as an n-component vector. However, many of the unique properties of cyclic codes are most apparent or easily proved when code words are represented by polynomials of degree $n - 1$. Thus, vector $\mathbf{x}$ and polynomial $x(t)$ will be used interchangeably, where

$$\mathbf{x} = (x_0, x_1, x_2, \ldots, x_{n-1}) \tag{4.5}$$

and

$$x(t) = x_0 + x_1 t + x_2 t^2 + \ldots + x_{n-1} t^{n-1} \tag{4.6}$$

In this book, the convention will be that the highest-degree term occurs earliest in time. From this assumption, it follows that multiplication by a power of t advances a vector in time, and the highest-degree terms (components) enter a processing element (encoder, multiplexer, *et cetera*) first. It should be pointed out that this particular time-ordering convention is not universal among authors, so the reader should be careful to interpret correctly what he or she is reading.

To show that constructing a cyclic code is not a trivial matter, consider the following example.

Example 4.10. Consider the (3, 2) linear code with generator matrix

$$G_1 = \begin{bmatrix} 1 & 1 & 0 \\ 0 & 0 & 1 \end{bmatrix}$$

Does G_1 generate a cyclic code? The rows of G_1 certainly are not cyclic shifts of each other. The third (and only other) nonzero vector of the code is 111, which is only a cyclic shift of itself. How about $G_2 = \begin{bmatrix} 1 & 0 & 0 \\ 0 & 1 & 0 \end{bmatrix}$? Either row of G_2 is a shift of the other, but the other shift, 001, cannot be constructed from G_2. Another possibility for a generator matrix is

$$G_3 = \begin{bmatrix} 1 & 1 & 0 \\ 0 & 1 & 1 \end{bmatrix}$$

The sum of the row vectors in G_3 is 101, which, together with the rows of G_3, yields a set of cyclic shifts of a single vector. (In all cases, the vector 000 satisfies the cyclic property.)

To understand why G_3 succeeded where G_1 and G_2 failed in generating a cyclic code, we must look at the results of the cyclic property on the code polynomials. These polynomials are tabulated here:

G_1	G_2	G_3
$t^2 + t$	t^2	$t^2 + t$
1	t	$t + 1$

A look at G_2 and G_3 and their corresponding code polynomials suggests that a single shift of a vector to the right corresponds to multiplying its corresponding polynomial by t. If this is true, then for G_3 the rightward cyclic shift of 110 should give 101, corresponding to $t^2 + t^3$. But this last polynomial has degree greater than $n - 1 = 2$. What can be done? It turns out that all polynomials must be reduced modulo $t^n - 1$ (i.e., $t^n + 1$ over GF(2)).

Therefore, because $n = 3$, you can deal with $t^2 + t^3$ in the following manner:

$$t^2 + t^3 = (t^2 + 1) + (t^3 + 1) = 1 + t^2 \bmod(t^3 + 1)$$

and $1 + t^2$ corresponds to the vector 101. The next shift would give

$$t + t^3 = (t + 1) + (t^3 + 1) = 1 + t \bmod(t^3 + 1)$$

which is again the second row of G_3. Finally, consider $1 + t + t^2$ (which is not in the code generated by G_3). The rightward shift of vector 111 corresponds to

$$t + t^2 + t^3 = (1 + t + t^2) + (t^3 + 1) = 1 + t + t^2 \quad \bmod(t^3 + 1)$$

Note that in simple cases this method of reducing modulo $t^n + 1$ is quicker than long division of the given polynomial by $t^n + 1$.

Concentrating on the cyclic code defined by matrix G_3, note that the lowest-degree polynomial occurring is $t + 1$. In general, the monic polynomial of lowest degree that is a code word is called the *generator polynomial* $g(t)$ of the cyclic code. A requirement on $g(t)$ is that $g_0 = g_{n-k} = 1$. It is easy to show (see Problems) that all polynomial multiples of $g(t)$, reduced modulus $t^n - 1$, are code words of the same cyclic code, that $g(t)$ must divide $t^n - 1$, and that the degree of $g(t)$ is $n - k$.

Example 4.11. You have already seen that the multiples of $g(t)$, mod$t^n - 1$, are code words for the code defined by G_3. Does $g(t) = t + 1$ divide $t^n - 1 = t^3 + 1$? Yes, because $t^3 + 1 = (t + 1)(t^2 + t + 1)$ over GF(2). It is also clear that the degree of $g(t)$ is $n - k = 3 - 2 = 1$.

With the properties just specified for $g(t)$, along with the correspondence stated at the start of this section between a code vector and its polynomial representation, there is sufficient information to write down the generator matrix for a cyclic code, given its generator polynomial. Because G for an (n, k) code must have exactly k linearly independent rows, take for these rows the vector corresponding to $g(t)$, along with its first $k - 1$ linear shifts. This gives a matrix G with k rows and $(n - k + 1) + (k - 1) = n$ columns, because the degree $n - k$ polynomial $g(t)$ has $n - k + 1$ terms, some possibly 0. Consequently, for

$$g(t) = g_0 + g_1 t + \ldots + g_{n-k-1} t^{n-k-1} + g_{n-k} t^{n-k}$$

there results the $k \times n$ matrix.

$$G = \begin{bmatrix} g_0 & g_1 & \cdots & g_{n-k-1} & g_{n-k} & & & & \\ & g_0 & g_1 & & \cdots & g_{n-k-1} & g_{n-k} & & \\ & & g_0 & & g_1 & \cdots & g_{n-k-1} & g_{n-k} & \\ & & \cdot & & & & & & \\ & & & \cdot & & & & & \\ & & & & \cdot & & & & \\ & & & & g_0 & g_1 & \cdots & g_{n-k-1} & g_{n-k} \end{bmatrix} \quad (4.7)$$

where the blank areas have all zeros as entries.

Example 4.12. Let $n = 7$, $k = 4$. Then $g(t)$ is of degree $n - k = 3$. Suppose $g(t) = 1 + t + t^3$ over GF(2). Then G has four rows and seven columns and is given by

$$G = \begin{bmatrix} 1 & 1 & 0 & 1 & 0 & 0 & 0 \\ 0 & 1 & 1 & 0 & 1 & 0 & 0 \\ 0 & 0 & 1 & 1 & 0 & 1 & 0 \\ 0 & 0 & 0 & 1 & 1 & 0 & 1 \end{bmatrix}$$

Any generator matrix for a cyclic code constructed in the manner of Example 4.12 can be transformed by row operations into a matrix for a systematic code. (See Section A.4 in the Appendix.)

Example 4.13. The matrix of Example 4.12 becomes successively

$$\begin{bmatrix} 1 & 1 & 0 & 1 & 0 & 0 & 0 \\ 0 & 1 & 1 & 0 & 1 & 0 & 0 \\ 0 & 0 & 1 & 1 & 0 & 1 & 0 \\ 0 & 1 & 1 & 1 & 0 & 0 & 1 \end{bmatrix}, \begin{bmatrix} 1 & 1 & 0 & 1 & 0 & 0 & 0 \\ 0 & 1 & 1 & 0 & 1 & 0 & 0 \\ 1 & 1 & 1 & 0 & 0 & 1 & 0 \\ 1 & 0 & 1 & 0 & 0 & 0 & 1 \end{bmatrix} = G_s$$

The matrix G_s represents the same code in systematic form with the last four digits as information digits.

In the polynomial representation of this code, the rows of the final systematic form G_s correspond to polynomials obtained by successively dividing information bits $t^3 = t^{n-k}$, t^4, t^5, and $t^6 = t^{n-1}$ by $g(t)$. We add on the remainder from each division operation to form the following polynomial:

$$t^i = q_i(t)g(t) + r_i(t) \quad i = n - k, \ldots, n - 1$$

where $r_i(t)$ has degree $n - k - 1$ or less. Rearrange the polynomial to obtain

$$q_i(t)g(t) = r_i(t) + t^i$$

and then write the coefficients of $r_i(t) + t^i$ as a row of G_s (with coefficients of increasing powers of t running from left to right). Thus, for example, you can verify that

$$t^6 = (t^3 + t + 1)g(t) + t^2 + 1$$

so that

$$(t^2 + t + 1)g(t) = 1 + t^2 + t^6$$

corresponding to the last row, 1010001, of G_s.

Corresponding to the H matrix is the check polynomial $h(t)$, which satisfies

$$g(t)h(t) = t^n - 1 \tag{4.8}$$

4.5 ENCODING OF CYCLIC CODES

One of the attractive features of cyclic codes is the ease with which encoding can be implemented through the use of linear shift registers. The shift register implements either the degree $-(n - k)$ generator polynomial $g(t)$ or the degree $-k$ check polynomial $h(t)$. The number of stages in the shift register is equal to the degree of the polynomial being implemented. Thus, the choice of implementation is frequently made on the basis of whether $n - k$ or k is the smaller number.

For binary codes, coefficient values of 0 and 1 are implemented by an open circuit and a short circuit, respectively. For codes over GF(q) with $q > 2$, the circuitry becomes slightly more complicated because multiplication of elements of GF(q) must be performed. An alternative to hard-wired logic would be a programmable microprocessor.

The encoding process will be illustrated with the (7, 4) cyclic code of Examples 4.12 and 4.13. This code has $g(t) = t^3 + t + 1$ and $h(t) = (t^n - 1)/g(t) = t^4 + t^2 + t + 1$.

In the shift register diagrams, the contents of each stage appear at the output of that stage. Assume that the shift register stages have all been set to zero prior to the start of a new encoding operation.

The shift register of Figure 4.1 will perform encoding based on $g(t)$ and will produce a systematic code. Initially, the gate is enabled and switch S is in position 1. This allows the four information digits to go to the modulator or channel while simultaneously entering the shift register to compute the check digits. After this entrance has taken place (requiring k shifts or clock pulses), the gate is disabled. S moves to position 2 and the $n - k$ check digits, which by that time occupy the shift register, are shifted out to the modulator or channel. This implementation exploits the fact that the information portion of the code word can be written using the Euclid division algorithm as $x_{n-1}t^{n-1} + x_{n-2}t^{n-2} + \ldots + x_{n-k}t^{n-k} = q(t)g(t) + r(t)$. Here $r(t)$ must be of degree at least one less than $n - k$, the degree of $g(t)$. Because of the cyclic nature of the code, $q(t)g(t)$ is a code word and so, therefore, is $x_{n-1}t^{n-1} + \ldots + x_{n-k}t^{n-k} - r(t)$. However, because the coefficients $x_{n-1}, \ldots, x_{n-k}$ are the information digits, the coefficients of $r(t)$ must be the check digits. Division of $x_{n-1}t^{n-1} + \ldots + x_{n-k}t^{n-k}$ by $g(t)$ is carried out by the circuit of Figure 4.1 and gives $r(t)$ as a remainder.

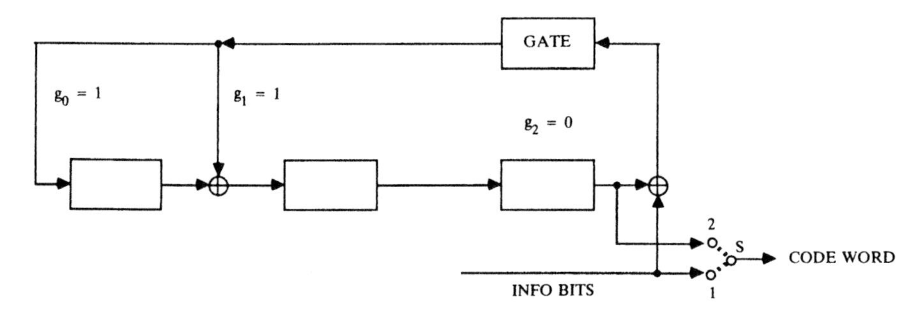

Figure 4.1 $(n - k)$-stage encoder for (7, 4) cyclic code.

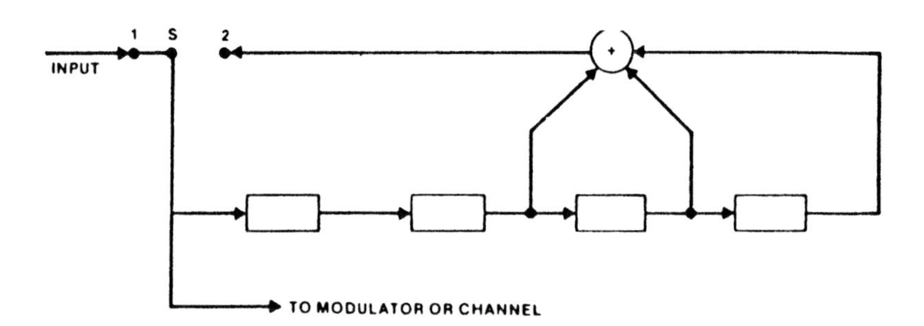

Figure 4.2 k-stage encoder for (7, 4) cyclic code.

The encoder of Figure 4.2 corresponds to $h(t)$. Initially, switch S is in position 1, where it remains until the four information digits are fed into the register and also to the modulator or channel. The switch then moves to position 2 and the shift register is stepped three times to produce the check digits. This method of encoding is based on the following chain of reasoning: if $x(t) = b(t)g(t)$ is a code word, then $x(t)h(t) = b(t)g(t)h(t) = b(t)(t^n - 1) = t^n b(t) - b(t)$, since $g(t)h(t) = t^n - 1$. Now, the degree of $x(t)h(t)$, deg(xh), is deg(x) + deg(h) = $(n - 1) + k = n + k - 1$, and deg(gh) = n. Therefore, deg(b) = deg(xh) $- n = k - 1$. In view of this, it follows that $t^n b(t)$ has only terms of degree $n, n + 1, \ldots, n + k - 1$, while $b(t)$ has terms of degree $0, 1, \ldots, k - 1$. Thus, $x(t)h(t)$ has no terms of degree $k, k + 1, \ldots, n - 1$, and this fact can be used to derive $n - k$ relationships of the form

$$\sum_{i=0}^{k} h_i x_{\ell - i} = 0 \quad k \leq \ell \leq n - 1 \tag{4.9}$$

which can be rewritten as explicit recurrence relationships for the check digits $x_0, x_1, \ldots x_{n-k-1}$ in terms of the information digits $x_{n-k}, \ldots, x_{n-1}$. The result is

$$x_{\ell - k} = - \sum_{i=0}^{k-1} h_i x_{\ell - i} \quad k \leq \ell \leq n - 1 \tag{4.10}$$

where the minus sign becomes plus in GF(2) and where ℓ must take on descending values $n - 1, n - 2, \ldots,$ to make use of the recurrence relation.

Chapter 5
Bose-Chaudhuri-Hocquenghem and Reed-Solomon Codes

5.1 INTRODUCTION

In this chapter, the more advanced algebraic concepts and the background in cyclic codes presented in Chapter 4 will be used to develop two classes of cyclic codes. One, the Bose-Chaudhuri-Hocquenghem (BCH) codes, is the most powerful known class of binary block codes for correcting random errors ([5.1], [5.2], [5.3]). The BCH codes generalize the single-error-correcting codes to the case of multiple, randomly distributed errors. (Contrast this situation with that of correcting clustered multiple errors, i.e., a burst.) The other class, the Reed-Solomon (RS) codes, is considered the best set of block codes for dealing with multiple bursts of errors in a code word. Encoding for BCH codes has always been easy to implement because of their cyclic code properties. Decoding was greatly simplified, both conceptually and in practice, by the contributions of Berlekamp [5.4] and Massey [5.5].

Initially, the RS code hardware implementation was considered cumbersome because of the nonbinary finite-field arithmetic required for manipulating the higher-order symbols of these codes. A number of low-cost implementations became available in the late 1970s. For example, Carhoun, Roome, and Palo [5.6, 5.7] reported on the use of charge-coupled devices. Integrated circuits are now being used (for example, in the Joint Tactical Information Distribution System (JTIDS) encoder-decoder).

5.2 DEFINITION, CONSTRUCTION, AND PROPERTIES OF BCH CODES

A BCH code, like other linear codes, can be defined in terms of its parity-check matrix H. Because it is a cyclic code, it can also be defined by its generator polynomial $g(t)$, and hence, its generator matrix G. The discussion here will address only the second of these approaches in view of the tabulations of BCH codes in terms of generators. Fortunately, in practice you generally do not need to start from scratch each time you desire a BCH code. Extensive tables of code generators are available and one will be given later in this chapter. As with all cyclic codes, the matrices G and H uniquely define the generator polynomial $g(t)$ and the check polynomial $h(t)$, respectively. Similarly, given n, $g(t)$ and $h(t)$ uniquely determine G and H, respectively. Finally, the relationships

$$GH = \Phi \tag{5.1a}$$

$$g(t)h(t) = t^n - 1 \tag{5.1b}$$

hold for BCH codes just as they do for all block codes (in the case of (5.1a)) and for all cyclic codes (as in (5.1b)). Here, as before, Φ is an n x $(n - k)$ all-zero matrix.

A BCH code will be defined here by making use of the fact that BCH codes are cyclic codes with particular conditions on the generator polynomial $g(t)$. As in Chapter 4, $g(t)$ is a monic polynomial of degree $r \geq 1$ with coefficients over code symbol field GF(q), and n is the smallest integer such that $g(t)$ divides $t^n - 1$. The additional condition on $g(t)$ in forming BCH codes is that $g(t)$ be the polynomial of lowest degree having $\alpha, \alpha^2, \ldots, \alpha^{d_0 - 1}$ as roots, and $d_0 \geq 2$ satisfies the condition that the $d_0 - 1$ numbers $\alpha, \ldots, \alpha^{d_0 - 1}$ are distinct roots of $g(t)$.

Therefore, a polynomial $g(t)$ generates a BCH code if and only if the properties just stated hold. Only binary BCH codes are used in practice, whereas RS codes are the choice for nonbinary systems. Therefore, this initial discussion will deal only with binary BCH codes, which result if $q = 2$. It is usual to let α be a primitive element of GF(2^m). Some BCH codes also result when α is not a primitive element, but this type will not be discussed here. The reader is referred to Lin and Costello [5.8].

As it turns out, the order $2^m - 1$ of α is equal to the code word length n in bits. The parameter d_0 is a lower bound to the actual minimum distance d of the code and is usually referred to as the designed distance. Note that with α as a primitive element of GF(2^m), the definition of d_0

implies that $d_0 - 1 \leqslant r$. If t is the actual minimum error-correcting capability of the code, then the relationships $d \geqslant 2t + 1$ and $d \leqslant n$ combine to give

$$2t + 1 \leqslant 2^m - 1$$

or (5.2)

$$t \leqslant 2^{m-1} - 1$$

Thus, for example, for $m = 4$, $n = 15$, the designed minimum error-correcting capability cannot exceed 7 (which is achieved for the code the only words of which are the all-zero and all-one words).

It may be helpful to work out what is actually a trivial case. In Example 5.1, it is obvious that there can be only two code words: 000 and 111.

Example 5.1. Let $m = 2$, $d = d_0 = 3$. Then $n = 2^2 - 1 = 3 = d$, $d_0 - 1 = 2$, and α and α^2 are distinct roots of $g(t)$, which must divide $t^3 - 1$ (or $t^3 + 1$, because coefficients are in GF(2)). Furthermore, because α is a primitive element of GF(2^2), it follows that $\alpha^3 = 1$. Thus $(t - \alpha)(t - \alpha^2) = t^2 - (\alpha + \alpha^2)t + 1 = g(t)$. Because the only first-degree polynomial that divides $t^3 + 1$ is $t + 1$, then $g(t)$ must be $(t^3 + 1)/(t + 1)$ or $t^2 + t + 1$. Thus, $\alpha + \alpha^2 = 1$. You can obtain these results even more easily by noting that $t^2 + t + 1$ is the only primitive second-degree polynomial over GF(2).

The theory of cyclic codes developed in the last chapter states that the degree of $g(t)$, which was found to be 2, is equal to $n - k$, so that $k = n - 2 = 1$. Furthermore, because generator matrix G has $n = 3$ columns, the coefficients of $g(t)$ fill the row. Thus, the generator matrix is simply $G = [1\ 1\ 1]$, because the coefficients of $g(t)$ and 0s (if necessary) make up any row of G. The code then consists of the vectors (0, 0, 0) and (1, 1, 1).

This example was intentionally worked in great and perhaps unnecessary detail to give you further practice and some insight in finite-field manipulations. (In particular, the rapid and unique determination of $g(t)$ actually eliminated the need to develop details of GF(2^2) because $n = 3$ and $g(t)$ turned out to have three terms.)

The next example is virtually a classic example of BCH code generation; it was used in 1960 by Bose and Ray-Chaudhuri [5.1] in their original paper and subsequently by Gallager [3.1] and by Peterson and

Weldon [3.2]. Note that the technique followed in Example 5.1 of calculating $g(t)$ as a product of linear factors, each having as a root a power of α, rapidly becomes difficult if we increase d_0. Perhaps the best way is to use a result stated earlier for Galois fields, in combination with tables of irreducible polynomials over GF(2) that have been developed by various workers. The Galois field result is this: if α is a primitive element of $GF(q^m)$, and $f(t)$ is the (mth degree)minimal polynomial of α, all elements of $GF(q^m)$ can be represented uniquely as polynomials in α by making use of the fact that $f(\alpha) = 0$. The important word here is "all." The tables of polynomials referred to are most readily available in Appendix C of Peterson and Weldon [3.2].

Example 5.2. Let $q = 2$, $m = 4$, $m_0 = 1$, $d_0 = 5$. Thus $n = 2^4 - 1 = 15$. Since $m_0 + d_0 - 2 = 4$, we must find minimal polynomials for α, α^2, α^3, and α^4. Because α, α^2, and α^4 all have the same minimal polynomial, we need to find minimal polynomials for α and α^3 only. Referring to the referenced tables in Peterson and Weldon [3.2], we find the following entries for irreducible polynomials of degree $m = 4$: 1 23F and 3 37D. (The letters shown are not important for this problem. The interested reader will find an explanation in the referenced appendix.) In "1 23F", the 1 indicates that it is α^1 for which 23 is the octal representation of the binary coefficients (highest degree term at the left) of the polynomial. Because 23 octal corresponds to 10 011 binary, this means that $t^4 + t + 1$ is the minimal polynomial of α. Similarly, the minimal polynomial of α^3 is $t^4 + t^3 + t^2 + t + 1$, corresponding to the table entry "3 37D." Thus, as required by the condition on $g(t)$ stated in the definition of a BCH code, $g(t)$ is the least common multiple of these two minimal polynomials:

$$\begin{aligned} g(t) &= (t^4 + t + 1)(t^4 + t^3 + t^2 + t + 1) \\ &= t^8 + t^7 + t^6 + t^4 + 1 \end{aligned} \tag{5.3}$$

which, in view of the cyclic nature of BCH codes, gives

$$G = \begin{bmatrix} 1&1&1&0&1&0&0&0&1&0&0&0&0&0&0 \\ 0&1&1&1&0&1&0&0&0&1&0&0&0&0&0 \\ 0&0&1&1&1&0&1&0&0&0&1&0&0&0&0 \\ 0&0&0&1&1&1&0&1&0&0&0&1&0&0&0 \\ 0&0&0&0&1&1&1&0&1&0&0&0&1&0&0 \\ 0&0&0&0&0&1&1&1&0&1&0&0&0&1&0 \\ 0&0&0&0&0&0&1&1&1&0&1&0&0&0&1 \end{bmatrix} \tag{5.4}$$

The theory and examples presented above should impart a basic understanding of BCH code structure and generation. In many cases, even the brief labor of Example 5.2 will not be necessary. For example, Stenbit [5.9] published a table of generators for all primitive BCH codes up to length 255. You should certainly consult this table, reproduced here as Table 5.1, when you want a BCH code. Coefficients are again in octal, right justified arrangement so that the zero-degree coefficient is at the extreme right.

Example 5.3. The code just found in Example 5.2 can be read directly from Table 5.1 as 721 for the (15, 7), two-error-correcting code.

5.3 DECODING

Many authors, including Gallager [3.1], Berlekamp [5.4], Lin [5.10], Peterson and Weldon [3.2], Chien [5.11], and McEliece [5.12], have already provided detailed descriptions of the decoding of BCH codes. Therefore, we shall not attempt either to repeat or to break new ground but rather, we shall only sketch out the steps involved. While encoding of BCH codes can be performed exactly as was described in Chapter 4 for any cyclic code, decoding is done in a manner that computes directly the locations and values of the individual symbol errors, in contrast to the storage-consuming table look-up associated with coset leaders and to the exhaustive calculation of the maximum likelihood estimate of the transmitted code word.

The decoding proceeds in four main steps:

1. Calculation of syndrome $s = (s_1, s_2, \ldots, s_{n-k})$;
2. Formation of the error-locating polynomial based on the components of the syndrome;
3. Calculation of the error locations;
4. Calculation of the $[(d - 1)/2]$ or fewer error values ($[x]$ denotes the largest integer equal to or less than x).

Step 4 is obviously unnecessary in the case of a binary code. Of the remaining steps, step 2 requires by far the most explanation and analysis, although in terms of concept and implementation, it is no more difficult than any of the other steps. Results published by Berlekamp [5.4] and Massey [5.5] made this step much easier to understand and carry out.

Steps 1, 2, and 3 of the BCH decoding procedure will now be discussed in further detail.

Step 1. Calculate Syndrome

Assume a received vector $\mathbf{y} = \mathbf{x} + \mathbf{e}$ where, as before, $\mathbf{x}$ is the transmitted code word and $\mathbf{e}$ is an error vector. Then the syndrome vector

Table 5.1 *Generators of Primitive BCH Codes [5.9]

n	k	t	$g(x)$
7	4	1	13
15	11	1	23
	7	2	721
	5	3	2467
31	26	1	45
	21	2	3551
	16	3	107657
	11	5	5423325
	6	7	313365047
63	57	1	103
	51	2	12471
	45	3	1701317
	39	4	166623567
	36	5	1033500423
	30	6	157464165547
	24	7	17323260404441
	18	10	1363026512351725
	16	11	6331141367235453
	10	13	472622305527250155
	7	15	5231045543503271737
127	120	1	211
	113	2	41567
	106	3	11554743
	99	4	3447023271
	92	5	624730022327
	85	6	130704476322273
	78	7	26230002166130115
	71	9	6255010713253127753
	64	10	1206534025570773100045
	57	11	335265252505705053517721
	50	13	54446512523314012421501421
	43	14	17721772213651227521220574343
	36	15	3146074666522075044764574721735
	29	21	403114461367670603667530141176155
	22	23	123376070404722522435445626637647043
	15	27	22057042445604554770523013762217604353
	8	31	7047264052751030651476224271567733130217
255	247	1	435
	239	2	267543
	231	3	156720665
	223	4	75626641375
	215	5	23157564726421
	207	6	16176560567636227
	199	7	7633031270420722341
	191	8	2663470176115333714567
	187	9	52755313540001322236351
	179	10	22624710717340432416300455
255	171	11	15416214212342356077061630637
	163	12	7500415510075602551574724514601
	155	13	3757513005407665015722506464677633
	147	14	1642130173537165525304165305441011711
	139	15	461401732060175561570722730247453567445
	131	18	215713331471510151261250277442142024165471
	123	19	120614052242066003717210326516141226272506267
	115	21	6052666557210024726363640460027635255631347273 7
	107	22	2220577232206625631241730023534742017657475015444 1
	99	23	10656667253473174222741416201574332252411076432303431
	91	25	675026503032744417272363172473251107555076272072434456 1
	87	26	110136763414743236435231634307172046206722545273311721317
	79	27	66700035637657500020270344207366174621015326711766541342355
	71	29	2402471052064432151555417211233116320544425036255764322170603 5
	63	30	107544750551635443253152173577070036661117264552676136567025433 01
	55	31	731542520350110013301527530603205432541432675501055704442603547361 7
	47	42	2533542017062646563033041377406233175123334145446045005066024552543173
	45	43	15202056055234161131101346376423701563670024470762373033202157025051541
	37	45	513633025506700741417744724543753042073570617432343234764435473740304400 3
	29	47	302571553667307146552706401236137711534224232420117411406025475741040356 5037
	21	55	1256215257060332656001773153607612103227341405653074542521153121614466513473725
	13	59	464173200505256454442657371425006600433067744547656140317467721357026134460500547
	9	63	157260252174724632010310432553551346141623672120440745451127661155477055 61677516057

*Reprinted with permission from *IEEE Trans. Information Theory*, volume IT-10, no. 4, October 1964, p. 391 [5.9].

is found, as usual, from $\mathbf{s} = \mathbf{y}H^T$. The cyclic nature of BCH codes leads to an important additional interpretation of the syndrome computation. Because every code polynomial $x(t)$ has α^i as a zero (i.e., $x(\alpha^i) = 0$) for α a primitive element of GF(2^m) and $i = 1, 2, \ldots, 2t$, it follows that $X(\alpha^i) = x_0 + x_1\alpha^i + x_2(\alpha^i)^2 + \ldots + x_{n-1}(\alpha^i)^{n-1} = \mathbf{0}$, which can be written as the product of a row vector and a column vector as follows:

$$(x_0 x_1 x_2 \ldots x_{n-1}) \begin{bmatrix} \mathbf{1} \\ \alpha^i \\ (\alpha^i)^2 \\ \cdot \\ \cdot \\ \cdot \\ (\alpha^i)^{n-1} \end{bmatrix} = \mathbf{0} \tag{5.5}$$

where the values x_i are elements of GF(2), and $\mathbf{0}$ and $\mathbf{1}$ are the additive and multiplicative identities, respectively, in GF(2^m). Similarly, $\mathbf{s} = \mathbf{y}H^T$ gives rise to the polynomial form.

$$s_i = y(\alpha^i) = y_0 + y_1\alpha^i + y_2(\alpha^i)^2 + \ldots + y_{n-1}(\alpha^i)^{n-1} \tag{5.6}$$

where again $i = 1, 2, \ldots, 2t$. Thus, each component of $\mathbf{s}$ is an element of GF(2^m). Furthermore, since $\mathbf{x}H^T = \mathbf{0}$, $\mathbf{y}H^T = \mathbf{e}H^T = \mathbf{s}$ then just as for any other block code, and therefore $s_i = e(\alpha^i)$, where

$$e(t) = e_0 + e_1 t + e_2 t^2 + \ldots + e_{n-1}t^{n-1} \tag{5.7}$$

is the error polynomial.

Step 2. Form Error-Locating Polynomial

For simplicity of notation, suppose that, unknown to us, there are errors in positions j through $j + E - 1$ of the code word, and assume for now that $E \leq t$. Then $\mathbf{e} = (e_0\, e_1 \ldots e_{n-1})$ has $e_j = e_{j+1} = \ldots = e_{j+E-1} = 1$ and $e_0 = e_1 = \ldots = e_{j-1} = e_{j+E-1} = \ldots = e_{n-1} = 0$. This means that $e(t) = t^j + t^{j+1} + \ldots + t^{j+E-1}$ and therefore,

$$\begin{aligned} S_1 &= e(\alpha) = \alpha^j + \alpha^{j+1} + \ldots + \alpha^{j+E-1} \\ S_2 &= e(\alpha^2) = (\alpha^2)^j + (\alpha^2)^{j+1} + \ldots + (\alpha^{2j+E-1})^1 \\ &= (\alpha^j)^2 + (\alpha^{j+1})^2 + \ldots + (\alpha^{j+E-1})^2 \\ S_3 &= e(\alpha^3) = (\alpha^j)^3 + (\alpha^{j+1})^3 + \ldots + (\alpha^{j+E-1})^3 \end{aligned} \tag{5.8}$$

and finally,

$$S_{2t} = e(\alpha^{2t}) = (\alpha^j)^{2t} + (\alpha^{j+1})^{2t} + \ldots + (\alpha^{j+E-1})^{2t}$$

This is a system of $2t$ nonlinear equations in the E unknowns α^j, α^{j+1}, $\ldots$, α^{j+E-1}. (Bear in mind that the real unknowns here are the position-locating exponents j, $j + 1, \ldots, j + E^{-1}$, which would generally be an arbitrary subset of the integers 0 through $n - 1$.) Of course, the error pattern can be corrected only if there are no more than t errors (i.e., t nonzero terms in $e(t)$), giving at least twice as many equations as there are unknowns. In this situation, no unique solution to (5.8) exists. Of the many possible solutions, then, which one is "best"? The answer comes from the general theory of decoding: choose the maximum-likelihood solution, i.e., the error pattern containing the smallest number of 1s. You now have a criterion for identifying the best solution to (5.8) of all possible solutions. However, this knowledge does not make the actual computation any easier.

The method of solving (5.8) involves the substitution $\beta_1 = \alpha$, $\beta_2 = \alpha^2$, *et cetera,* to make the notation less cumbersome; introducing a new polynomial

$$\begin{aligned}\sigma(t) &= (1 + \beta_1 t)(1 + \beta_2 t) \ldots (1 + \beta_E t) \\ &= \sigma_0 + \sigma_1 t + \sigma_2 t^2 + \ldots + \sigma_E t^E = 0 \qquad (5.9)\end{aligned}$$

and observing that the zeros β_1^{-1}, β_2^{-1}, *et cetera,* of this *error-location polynomial* $\sigma(t)$ are the inverses of the powers of α in (5.8). Thus, instead of solving $2t$ nonlinear equations in E unknowns (the α^l), you solve a single polynomial equation of degree E. This equation has E roots from which you can find original powers of α directly and uniquely. The exponents, i.e., the error locations, can then be determined by a table look-up in $GF(2^m)$.

Step 3. Calculate Error Locations

The tough part is now finding the coefficients σ_l, which give an equation of minimum degree E because this is equivalent to finding a minimum-weight error vector and hence a maximum-likelihood solution to the original decoding problem. Another very useful relationship can be introduced to help with the computation of the coefficients σ_l. This relationship is a set of equations called Newton's identities, which relate σ_l to the syndrome components:

$$s_1 + \sigma_1 = 0$$

$$s_2 + \alpha_1 s_1 + 2\sigma_2 = 0$$

$$s_3 + \sigma_1 s_2 + \sigma_2 s_1 + 3\sigma_3 = 0$$

$$\vdots$$

$$s_E + \sigma_1 s_{E-1} + \ldots + \sigma_{E-1} s_1 + E\sigma_E = 0 \qquad (5.10)$$

and, if necessary to continue,

$$S_{E+1} + \sigma_1 s_E + \ldots + \sigma_{E-1} s_2 + \sigma_E s_1 = 0$$

Bear in mind that (5.10) can include at most $2t$ nontrivial equations. The chief advantage of using the Newton's identities is that they are linear in the unknown values σ_l expressed in terms of the known values s_1, $s_2, \ldots, s_{2t}$.

Because the major constraint is now that (5.9) be of minimum degree, solving (5.10) is not as simple as it might seem. The most common method of solution at this time involves an iterative procedure developed by Berlekamp. This procedure produces the coefficients σ_l to specify (5.9). Solution of (5.9) then gives the values of β. Because all the equations throughout this discussion have coefficients (and therefore roots) from $GF(2^m)$, the usual techniques of solving equations over the real or complex fields do not apply. The basis of most of the techniques for solving (5.9) is to substitute elements of $GF(2^m)$ into (5.9) until you find one that works.

A more detailed discussion of the computational techniques for solving (5.9) and (5.10) would be too complex for our purposes. Therefore, the reader is referred to the outstanding texts by Lin and Costello [5.8] or Michelson and Levesque [3.4] for complete details and examples.

5.4 PERFORMANCE OF BCH CODES

Each BCH code has its designed error-correcting capability specified as a parameter ($d_0 = 2t + 1$) of its construction. Alternatively, tabulations of codes (e.g., Appendix D of Peterson and Weldon [3.2] and Table 5.1) always specify t as one of the code parameters. It is well known, however, that most BCH codes can correct significant numbers of error patterns that exceed the designed error-correcting capability of the code. There are two reasons for this. The first reason is that the designed distance d_0 may be less than the actual minimum distance d_{min}. A look at Appendix D of Peterson and Weldon amply bears this out. The second reason is that certain error patterns with weight greater than the guaranteed, rather than designed, error-correcting capability are nonetheless correctable.

On the negative side, it has been shown that $d_0/n \rightarrow 0$ as n becomes large with rate k/n held fixed. Since it can also be shown that $d_{min} \leqslant 2\ d_0$ at all times, it follows that $d_{min}/n \rightarrow 0$ also. Thus, BCH codes show decreasing error-correcting power (relative to block length) as block length increases and rate is held fixed.

5.5 REED-SOLOMON CODES

The Reed-Solomon [5.13] codes may be viewed as a subclass of the BCH codes in which $m = 1$ in GF(q^m) and $q > 2$. In other words, the RS codes are nonbinary BCH codes having $n = q - 1$. A special property of these codes is that their (BCH) decoding algorithm has the symbol field identical to the error-locator field. In terms of minimum distances, the RS codes do as well as possible for a given n and k, namely

$$d_{min} = n - k + 1$$

Remember that these numbers all pertain to symbols from a higher-order alphabet ($q > 2$).

If $q = p^s$, where p is a prime and s is a positive integer, then the RS codes can be a class of codes first considered by Gorenstein and Zierler [5.14]. In particular, if $p = 2$, the result is the Zierler codes, which will be discussed in Chapter 10 as a burst-control technique. The case $p = 2$ can be implemented in terms of binary logic elements. Because $n = q - 1$ for the RS codes, we also have $n = p^s - 1$ and $k = n - d + 1 = p^s - d$, all in q-ary symbols. Finally, because $d = 2t + 1$, where t is the (q-ary) error-correcting capability of the code, we have $k = p^s - 2t - 1$. In summary, we can construct an $(n, k) = (p^s - 1, p^s - 2t - 1)$ code capable of correcting any combination of t errors and e erasures satisfying $2t + e \leqslant n - k$ in symbols from GF(p^s). Here now is the origin of the multiple-burst-correcting capability of the RS (and Zierler) codes. Each symbol from GF(2^s) can be represented as a set of s binary digits. Thus, $2t$ q-ary symbols are equivalent to $2ts$ binary digits. In terms of binary digits, then, we have a code of length $s(2^s - 1)$ digits, of which $s(2^s - 2t - 1)$ digits correspond to information.

RS decoders are now commercially available. Their initial development was spurred by military tactical communication requirements.

Example 5.4. It is not unusual to take $s = 4$. Thus $n = 2^4 - 1 = 15$. If $t = 2$, then $k = 15 - 2 \times 2 = 11$. In terms of binary digits, this (60, 44) code can correct up to eight errors, provided they occur in bursts covering no more than two 16-ary symbols. (Each symbol will be one of the 4-tuples 0000, 0001, . . . , 1111.)

Because an RS code is, in one sense, a special case of the BCH codes, you can construct its generator polynomial in the same way as you construct the $g(t)$ for any other BCH code. In this case, however, the roots of the minimal polynomial $g(t)$ are all in the $g(t)$ coefficient field, GF(q). That is, $g(t)$ can be factored completely into *linear* factors. Thus, the BCH code construction rule yields

$$g(t) = (t - \alpha)(t - \alpha^2) \ldots (t - \alpha^{d_0 - 1}) \tag{5.5}$$

for the generator polynomial of a RS code, where the powers of α are elements of GF(q).

Unlike binary BCH codes, RS codes have not been tabulated as a special class, so there is no table of RS codes similar to Table 5.1 for BCH codes. Table 5.2, however, shows most of the useful parameters for application of a small number of RS codes. The quantity nm in the last column of Table 5.2 is the number of bits (as opposed to symbols) in a code word.

You really do not need detailed data for RS codes like those provided in Table 5.1. The code parameters (as in Table 5.2) are easily computed, and the generator polynomial is found as a product of linear factors (as in (5.5).

This chapter concludes with two examples of RS code construction. In these two examples, the RS code is defined over GF(2^3), so $n = q - 1 = 7$. Because q is thus a power of 2, code symbols can conveniently be represented by groups of three binary digits, which can easily be implemented in standard digital logic. (Note that q could take on any value; a good illustration of a case in which $q \neq 2^m$ occurs on page 187 of Michelson and Levesque [3.4]. However, the ease of implementation is not as great when q is not a power of 2.)

Example 5.5. First consider an RS code over GF(2^3), which corrects 3 errors. (Note that there are 3 bits per symbol here, so 3 symbol errors can encompass anywhere from 3 to 9 bit errors, depending on the distribution

Table 5.2 RS Code Parameters

$q = 2^m \quad n = q - 1 \quad n - k = 2t \quad d_{\min} = 2t + 1$
(All values are expressed in m-bit symbols, except for m and nm.)

m	n	t	k	$d_{\min}$	nm
2	3	1	1	3	6
3	7	1	5	3	21
		2	3	5	
		3	1	7	
4	15	1	13	3	60
		2	11	5	
		3	9	7	
		4	7	9	
		5	5	11	
		6	3	13	
		7	1	15	
5	31	1	29	3	155
		5	21	11	
		8	15	17	
8	255	5	245	11	2040
		15	225	31	
		50	155	101	

of the bit errors.) Then $d = 7$, so that $k = n - d + 1 = 1$. Note that this still results in a code having $q^k = 8$ distinct code words. The exponent $d - 1 = 6$, so that

$$g(t) = (t - \alpha)(t - \alpha^2)(t - \alpha^3)(t - \alpha^4)(t - \alpha^5)(t - \alpha^6)$$

Here α is a primitive element of GF(2^3), which is displayed in Table 4.1. The multiplication of the six linear factors to form the polynomial $g(t)$ is most easily performed by pairs and by using the form α^i for elements of GF(2^3). For instance,

$$\begin{aligned} g(t) &= [t^2 - (\alpha + \alpha^2)t + \alpha^3][t^2 + (\alpha^3 + \alpha^4)t + \alpha^7] \\ &\quad \cdot [t^2 - (\alpha^5 + \alpha^6)t + \alpha^{11}] \\ &= (t^2 - \alpha^6 t + \alpha^3)(t^2 - \alpha t + 1)(t^2 - \alpha^3 t + \alpha^4) \end{aligned}$$

Here, reductions of sums have been carried out based on Table 4.1, such as $\alpha^5 + \alpha^6 = (1 + \alpha) + (\alpha + \alpha^2) = \alpha^3$. Also, because $\alpha^7 = 1$, it follows that $\alpha^8 = \alpha$. Other reductions are done in a similar way.
The final result is

$$g(t) = t^6 + t^5 + t^4 + t^3 + t^2 + t + 1$$

from which

$$G = [1 \quad 1 \quad 1 \quad 1 \quad 1 \quad 1 \quad 1]$$

Note that G has $k = 1$ row and $n = 7$ columns of elements of GF(2^3). A code word is still given by $\mathbf{x} = uG$, which reduces here to $\mathbf{x} = \alpha^i G$, with $i = -\infty, 0, 1, \ldots, 6$; that is, the code consists entirely of scalar multiples of the single row vector that constitutes G.

An example involving less computation but a more interesting G matrix follows.

Example 5.6. Again take symbols from GF(2^3) but let $t = 1$ so that $d = 3$, $k = 5$, and $d - 1 = 2$. Therefore,

$$g(t) = (t - \alpha)(t - \alpha^2) = t^2 - \alpha^6 t + \alpha^3$$

where the coefficient of t has been reduced to a single term using Table 4.1.

In this case,

$$G = \begin{bmatrix} \alpha^3 & \alpha^6 & 1 & & & & \\ & \alpha^3 & \alpha^6 & 1 & & & \\ & & \alpha^3 & \alpha^6 & 1 & & \\ & & & \alpha^3 & \alpha^6 & 1 & \\ & & & & \alpha^3 & \alpha^6 & 1 \end{bmatrix}$$

where the blank entries are all 0s, and each entry is an element of GF(2^3).

Chapter 6
Convolutional Codes and Encoding

Recall that encoding for a block code involves some large number (roughly 10 to 100) of information bits per code word or block. In sharp contrast to this, you will see that convolutional encoding generally encodes data bits no more than four at a time and usually just one at a time. Encoding is performed with one or more shift registers (SRs), which compute check symbols as linear combinations of the data bits contained in the SRs. Just as for block codes, there are systematic and nonsystematic convolutional codes corresponding to the cases in which the information bits are transmitted unaltered or altered, respectively. Both binary and nonbinary codes will be discussed.

6.1 CONVOLUTIONAL ENCODING

Similar to the case of continuous functions of a continuous variable (for example, a time-dependent voltage and current), convolution of two digital sequences can be accomplished as follows:

1. Reverse one sequence.
2. Align the digit at one end (say the right) of this reversed sequence with the digit at the other (in this case the left) end of the other sequence.
3. Multiply the digits that are aligned, add the products formed for this alignment of sequences, and regard the sum of products just found as one element of the convolution of the two sequences.
4. Shift the two sequences one position relative to each other.
5. Repeat steps 3 and 4 until the two sequences have been shifted completely past each other.

In the case of binary functions and sequences, the product at each position is either 0 or 1, and summing takes place modulo 2. This will be the case throughout this discussion of convolutional codes.

Example 6.1. Suppose you want to find the convolution of the sequences 1101 and 10011. It is customary to indicate this operation by writing (1101)∗(10011). In this case, the sequence 10011 will be held fixed while reversing 1101. Thus,

				1	0	0	1	1
1	0	1	1					

To see how the calculation procedure works, look at the first four steps of the convolution. Assume all zeros preceding and following each of the sequences.

(1)

			1	0	0	1	1
1	0	1	1				

(2)

	1	0	0	1	1	
1	0	1	1			

(3)

	1	0	0	1	1
1	0	1	1		

(4)

1	0	0	1	1
1	0	1	1	

Thus, the results at each of the four steps shown are given by the following:

Step 1 $1 \cdot 1 = 1$
Step 2 $1 \cdot 1 + 0 \cdot 1 = 1$
Step 3 $1 \cdot 0 + 0 \cdot 1 + 0 \cdot 1 = 0$
Step 4 $1 \cdot 1 + 0 \cdot 0 + 0 \cdot 1 + 1 \cdot 1 = 0$

You should convince yourself that the complete convolution is given by (reading left to right with increasing time) 11000111.

In convolutional encoding, it is usual to think of the 1s of the fixed sequence as representing the taps of a shift register while the moving sequence corresponds to the data or information bits as they move through the shift register.

Encoding takes place by advancing the information bits one or a few at a time through one or more shift registers. Let n_0 channel symbols be the total output of the shift registers for every k_0 input symbol. Then the code rate is

$$R = k_0/n_0$$

Particularly common cases are $k_0 = 1$ and $n_0 = 2, 3$, or 4. Obviously, a bit can influence the shift register output only as long as it is in the shift register. Thus, we usually refer to K, the number of stages in the shift register, as the *constraint length* of the convolutional code. Some practitioners use the term "constraint length" to refer to the maximum number $n_A = n_0K$ of channel symbols influenced by any bit passing through the shift register. The constraint length is roughly analogous to block length for block codes. The relationship between SR tap connections and mathematical representations of a code will be presented in Section 6.3.

Figure 6.1 shows an encoder for a systematic convolutional code having $R = 1/3$, $K = 3$. This code, chosen for its simplicity, is nonetheless adequate to illustrate all properties of a typical convolutional encoder. For the encoder of Figure 6.1, data bits enter the shift register from the right, one at a time. In addition to being transmitted unchanged (top output line), each bit is added modulo 2 to the preceding bit (middle output line) and to the two preceding bits (bottom output line), giving a total of $n_0 =$ 3 channel symbols out via the commutator for every $k_0 = 1$ bit into the shift register.

A requirement for any encoder SR is that the first and last stages must each have at least one tap; otherwise, the stage could be omitted with no effect on the SR output.

Example 6.2. To see how the encoder SR of Figure 6.1 operates, look at "snapshots" of the SR as the successive information bits 1101 enter from the right. For simplicity, the information preceding and following this four-bit sequence consists entirely of zeros. Figure 6.2 shows the details of computing the first of the two check digits that follow each information bit into the channel. The SR contents along with all information and check digits that make up the encoder output are summarized in Table 6.1.

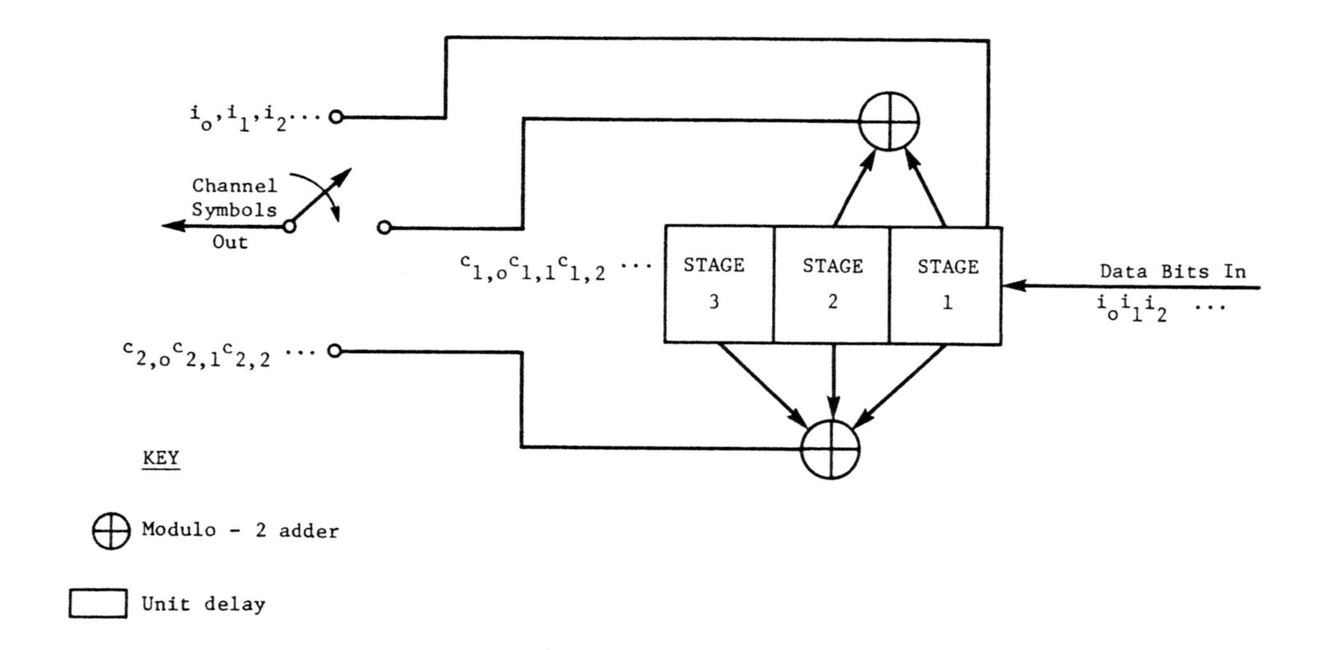

Figure 6.1 Convolutional encoder.

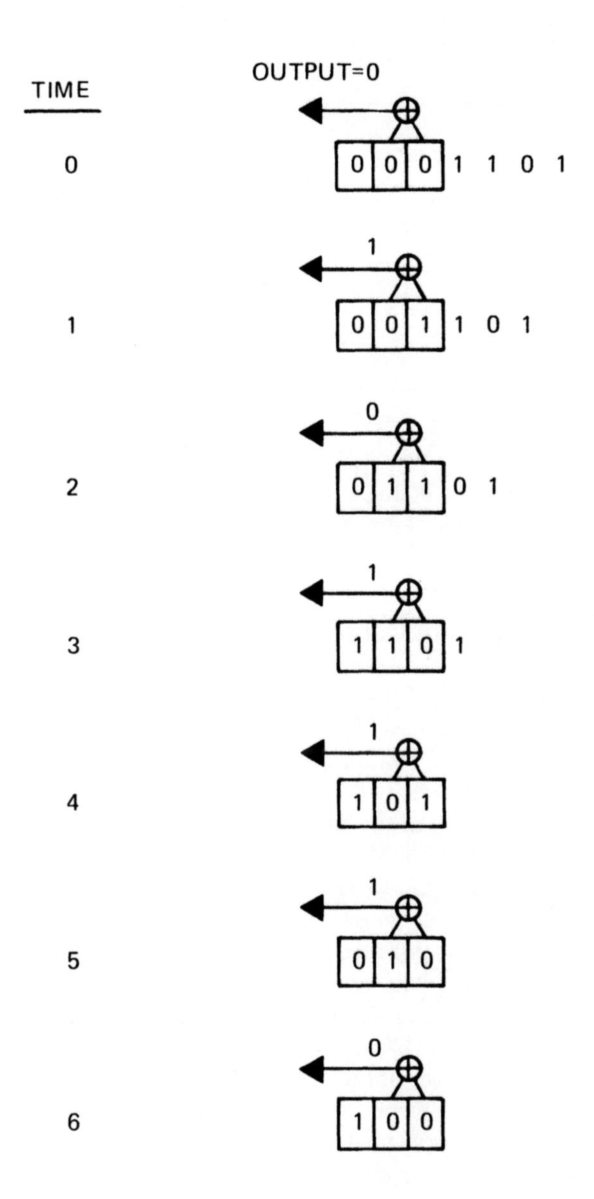

Figure 6.2 Generation of check digits by upper set of tap connections for encoder of figure 6.1.

Table 6.1

Time	*SR Contents*	*Output*
		$i\ c_1\ c_2$
0	0 0 0	0 0 0
1	0 0 1	1 1 1
2	0 1 1	1 0 0
3	1 1 0	0 1 0
4	1 0 1	1 1 0
5	0 1 0	0 1 1
6	1 0 0	0 0 1

Notice that the first of the three output digits at each discrete value of time is simply the latest input bit, in keeping with the systematic nature of the code.

6.2 GRAPHICAL REPRESENTATION OF CONVOLUTIONAL CODES

The possible sequences of channel symbols are conveniently displayed by a *tree*. The tree for the code of Section 6.1 is shown in Figure 6.3. Each branch of the tree represents the output corresponding to a particular set of three bits in the shift register; i.e., the two bits that were shifted left after encoding for the preceding bit time interval, and the new bit just introduced at the right-hand end of the shift register.

From left to right, the channel symbols appear on each branch in the order in which they are transmitted by the clockwise-rotating commutator of Figure 6.1. Each new data bit corresponds to a node of the tree. The number of possible branches emanating from a node of the tree is equal to the size of the input (or data, or information) alphabet; in this case, we are dealing with a binary alphabet, so that two branches leave every node. By convention, upper and lower branches from a node correspond to values of 0 and 1, respectively, for the data bits. A particular input sequence to the encoder corresponds to a unique path through the tree. For example, the data sequence 1101001 . . . (oldest digit at left) corresponds to the path shown with the heavy line in Figure 6.3.

Notice several properties of the branches of the tree. First, any succession of the branches shown can be generated by supplying the corresponding set of input bits to the encoder SR of Figure 6.1. Second, of the branches emanating from any node, all the digits on one branch are the binary

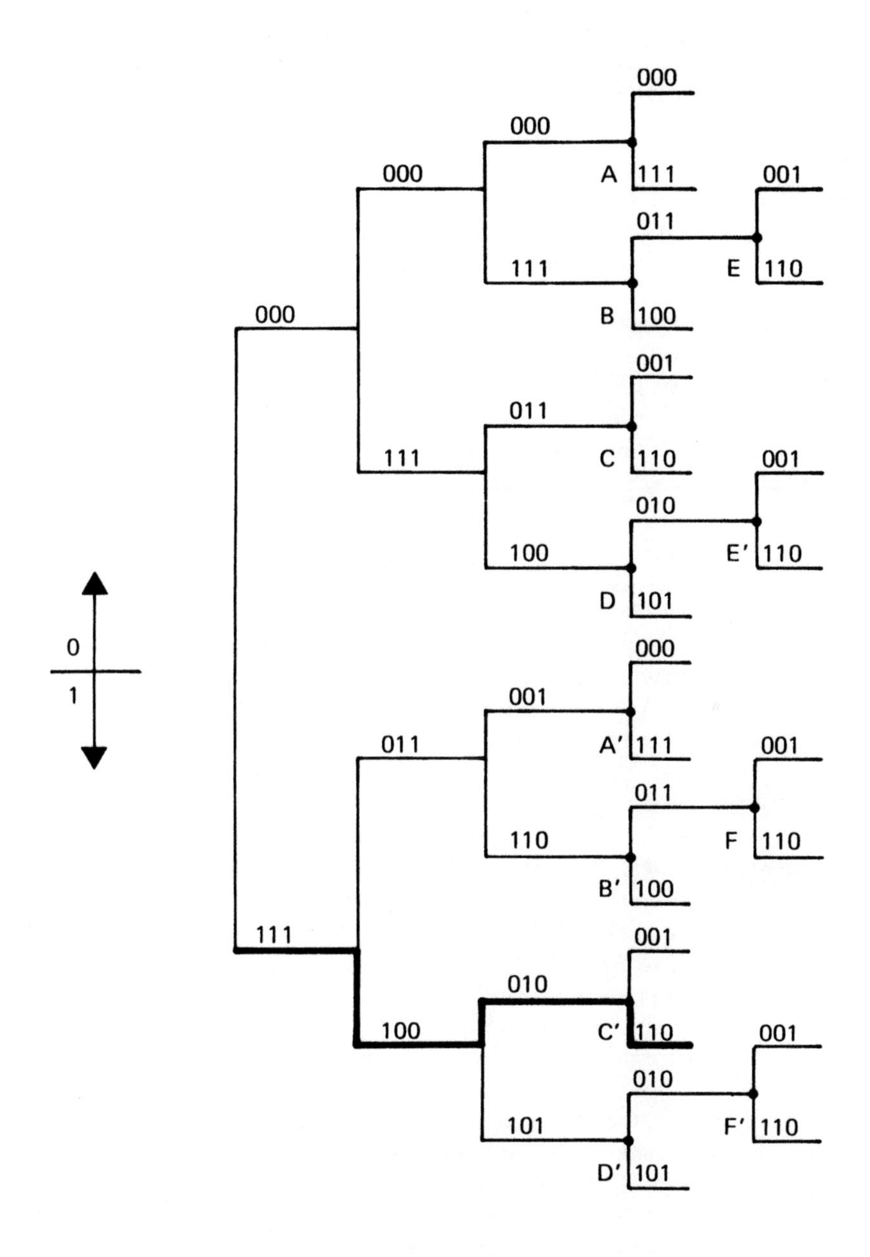

Figure 6.3 Code tree corresponding to encoder of Figure 6.1.

complement (1 instead of 0, 0 instead of 1) of the digits on the other branch. The reason for this is fairly clear: as long as the SR is such that each new input digit influences all output digits, changing the latest input to the SR changes all $n_0 - 1$ of the modulo -2 sums that determine the check digits.

Starting with the fourth branch of any sequence, identical sets of branches appear in the upper and lower halves of the tree, implying that certain pairs of nodes could be tied together and hence combined. This is true, for example, at the nodes marked A and A′ in Figure 6.3. The same is true of the portions beyond B and B′, C and C′, and D and D′. In view of this property of the code tree, we may view A and A′ as the same node. Similarly, each of the pairs B and B′, C and C′, and D and D′ reduces to a single node. Looking at the next level of nodes, observe that E and E′, F and F′, *et cetera,* may be tied together in a manner similar to that for A and A′, *et cetera.* Note that, with A and A′, . . . , D and D′ tied together, E already coincides with F, and E′ with F′. Thus, at each depth within the tree of one code constraint length, the number of branches can be halved without reducing the amount of information that can be obtained from the code tree diagram.

Combining nodes of the code tree in the manner just described results in a structure having a repetitive pattern, which is called a *trellis* because of its appearance. We use the terms *node* and *branch* for trellises exactly as in the case of the tree. In addition to displaying compactly the possible encoded sequences, the trellis will prove very useful in studying the details of the Viterbi decoding algorithm presented in Chapter 8. Because each new bit entering the encoder causes a bit to be "pushed out" at the other end, only the newest $K-1$ bits can influence the encoder output corresponding to the input bit about to enter the SR. We talk, therefore, of the 2^{K-1} possible *states* of the encoder at any time, along with a set of output symbols corresponding to each choice of input bit. On the trellis of Figure 6.4 (which corresponds to the encoder of Figure 6.1), each encoder state (having oldest bit on the left, newest on the right) is listed alongside the corresponding line of nodes in the trellis, while each set of output symbols appears on the branch connecting the state and its possible successor. As with the tree, the convention is to choose the upper and lower branches emanating from a node to correspond to the inputs 0 and 1, respectively. You should compare the SR states displayed in Figure 6.4 with the first digits of the SR contents in Table 6.1. Similarly, compare the outputs in Figure 6.4 and Example 6.2. The SR input is the last (right-hand) digit shown in the "SR Contents" column.

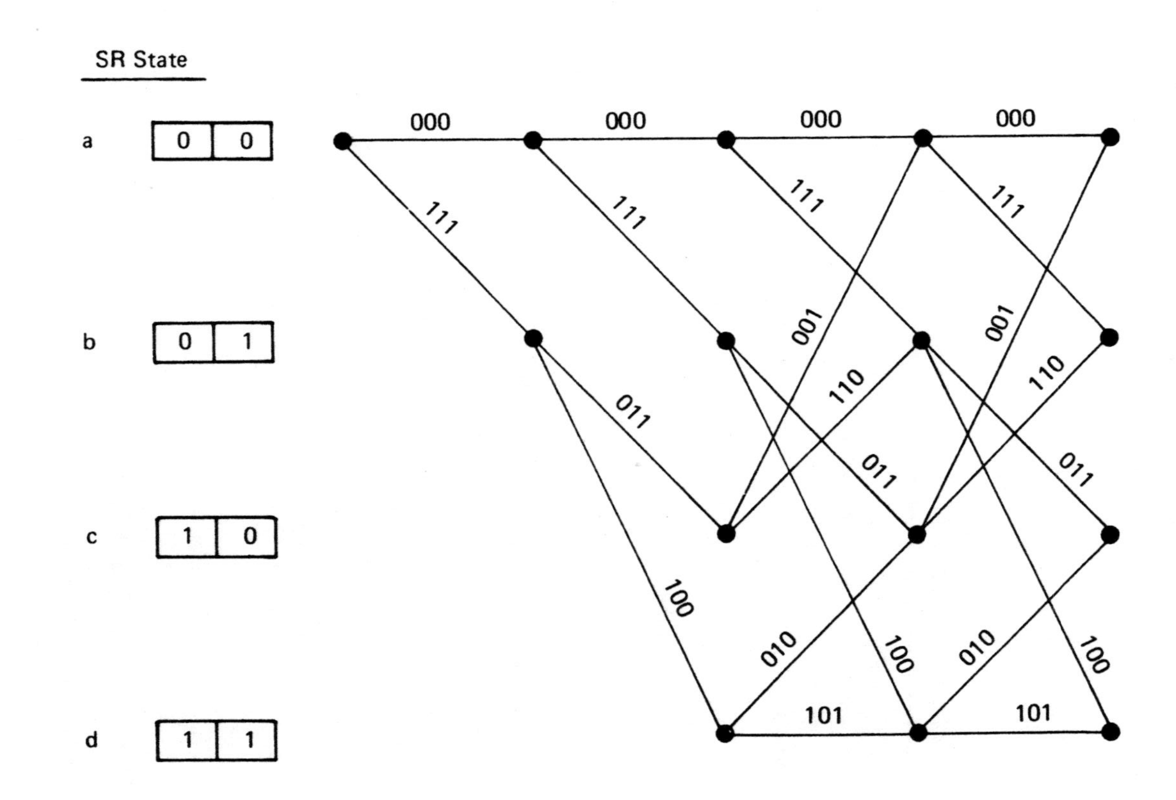

Figure 6.4 Trellis corresponding to the encoder of Figure 6.1.

The designation of encoder states in conjunction with the trellis leads to a third and final way of describing the operation of a convolutional encoder. This is the *state diagram,* a notation widely used in the study of discrete Markov processes and finite-state machines. In fact, a convolutional encoder is a finite-state machine; it has a defined set of states, a set of inputs, and a set of outputs, each of which depends on the state and the input. In addition, each state transition depends on the current state but not on any previous state; that is, the way in which the encoder arrived at its present state has no effect on future states.

Refer now to Figure 6.5, which corresponds to the code presented in Figures 6.3 and 6.4. Each node of the state diagram represents a unique state of the encoder. State transitions are indicated by directed line segments from one node to another. Note that for the 00 and 11 states, the encoder can remain in the same state, given the proper input. In all cases, each line segment is labeled with the encoder output for that state transition and input, with the input given in parentheses following the output. As before, the oldest bit is on the left in the state designation.

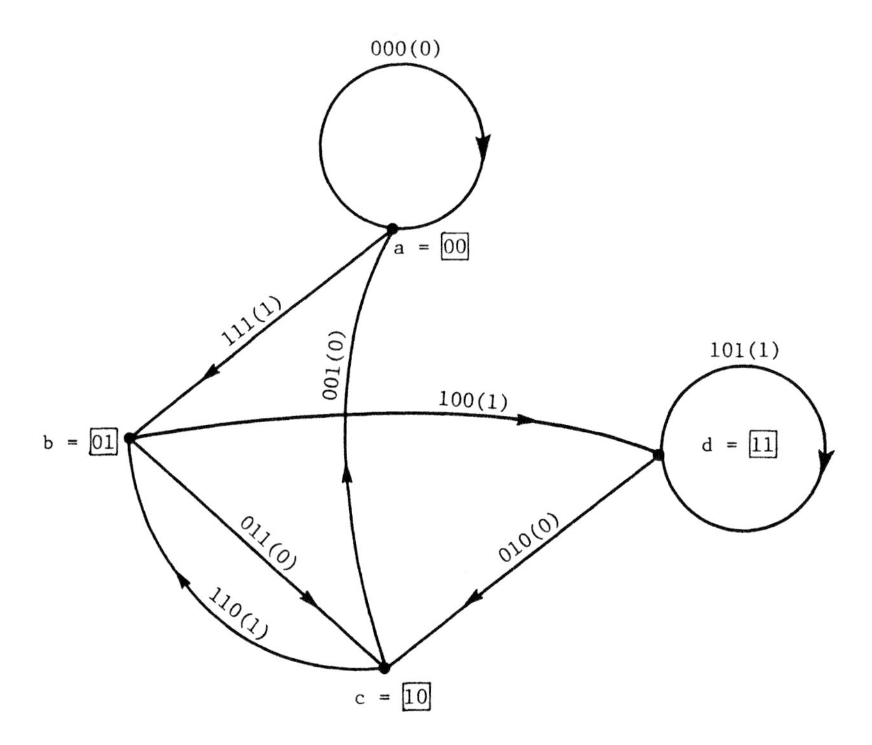

Figure 6.5 State diagram corresponding to the encoder of Figure 6.1.

6.3 ANALYTICAL REPRESENTATION OF CONVOLUTIONAL CODES

In contrast to the pictorial representations of convolutional codes given in Section 6.2, there are two very useful and widely used analytical representations. They are the delay-operator polynomial and the infinite (or semi-infinite) matrix.

In the first representation, both the encoder connections and the sequences to be encoded are represented as polynomials in the delay operator D. For example, the sequence 1101001, with oldest digit at the left, would be written in the delay operator notation (*transform* of the sequence) as

$$I(D) = 1 + D + D^3 + D^6 \tag{6.1}$$

Therefore, the exponent of the indeterminate D (sometimes incorrectly referred to as a dummy variable) is the number of time units of delay of that digit relative to the chosen time origin, which is usually taken to coincide with the first bit considered. More generally, the sequence $i_0\, i_1\, i_2\, i_3 \ldots$ has the transform $i_0 + i_1 D + i_2 D^2 + i_3 D^3 + \ldots$.

In representing shift register tap connections, an output of a stage to the appropriate modulo -2 adder corresponds to a coefficient of 1 in the polynomial, while no output corresponds to a 0. Thus, for upper and lower sets of taps, respectively, the SR of Figure 6.1 has these polynomial representations:

$$\begin{aligned} G_1(D) &= 1 + D \\ G_2(D) &= 1 + D + D^2 \end{aligned} \tag{6.2}$$

In this case, the lowest degree term corresponds to the stage encountered earliest by the entering bit stream. Following Massey, this discussion refers to polynomials that specify the SR tap connections as *generator polynomials.* Each such polynomial is the transform of what Wozencraft and Reiffen [6.1] referred to as a generator sequence.

Example 6.4. In contrast to the convolution operation, the use of transforms of the sequences is much more suited to computation with pencil and paper. For example, the convolution of Figure 6.2 corresponds to

$$\begin{aligned} (1 + D) \cdot (1 + D + D^3 + \ldots) &= 1 + (1 + 1)D + D^2 \\ + D^3 + D^4 + \ldots &= 1 + D^2 + D^3 + D^4 + \ldots \end{aligned}$$

Compare the final expression in this example with its counterpart in Example 6.2. Note that convolution of sequences corresponds to a multiplication of the respective transforms, just as in Fourier, Laplace, or z-transform theory.

The other major representation of convolutional codes and encoding uses semi-infinite vectors and matrices (so called because they have a definite beginning but no end). That is, the data vector is of the form:

$$\mathbf{u} = [u_1\ u_2\ u_3 \ldots]$$

where the elements of the vector **u** continue indefinitely to the right. Let the transmitted vector **x** be given by

$$\mathbf{x} = [x_{1,1}x_{1,2}x_{1,3}x_{2,1}x_{2,2}x_{2,3}x_{3,1} \ldots]$$

Example 6.5. Let us work out the form that the matrix G must have to give the transmitted vector $\mathbf{x} = \mathbf{u}\,G$.

Think in terms of the vector **u** entering the SR of Figure 6.1. For u_1 in stage 1 and 0s in the remaining two stages, the expressions for the first three transmitted symbols are

$$x_{1,1} = u_1$$

$$x_{1,2} = u_1$$

$$x_{1,3} = u_1$$

Now shift **u** one stage to the left. This gives

$$x_{2,1} = u_2$$

$$x_{2,2} = u_1 + u_2$$

$$x_{2,3} = u_1 + u_2$$

Finally, for u_1, u_2, and u_3 in the SR, observe that

$$x_{3,1} = u_3$$

$$x_{3,2} = u_3 + u_2$$

$$x_{3,3} = u_3 + u_2 + u_1$$

Note that this is the first time that all $K = 3$ stages of the SR have contained information bits; it can be thought of as a "steady-state" condition, while the preceding $(K - 1)$-step build-up is an "initial transient." Thus, in general, the SR steady-state output of Figure 6.1 is

$$x_{m,1} = u_m$$

$$x_{m,2} = u_m + u_{m-1}$$

$$x_{m,3} = u_m + u_{m-1} + u_{m-2}$$

The submatrix A_1 of G that gives this output must therefore satisfy

$$[u_{m-2} u_{m-1} u_m]\, A_1 = [u_m,\ u_m + u_{m-1},\ u_m + u_{m-1} + u_{m-2}]$$

A little thought (or grinding out the equation in terms of general elements of A_1) shows that

$$A_1 = \begin{bmatrix} 0 & 0 & 1 \\ 0 & 1 & 1 \\ 1 & 1 & 1 \end{bmatrix}$$

How can you handle the "initial transient"? As u_1, u_2, and u_3 are shifted into the SR, the initial shift (bringing in u_1) permits only the right-hand stage of the SR to influence the output. Next, the two right-hand stages influence the output. Finally, u_3 enters, and all three stages can influence the SR output. The submatrices $T_1, \ldots, T_{K-1} = T_2$, that give this "transient" behavior of the SR are

$$T_1 = \begin{bmatrix} 1 & 1 & 1 \\ 0 & 0 & 0 \\ 0 & 0 & 0 \end{bmatrix} \quad T_2 = \begin{bmatrix} 0 & 1 & 1 \\ 1 & 1 & 1 \\ 0 & 0 & 0 \end{bmatrix}$$

Thus, this analysis has shown that G is given by

$$G = \begin{bmatrix} [T_1]\,[T_2]\,[A_1] & 0\,0\,0 & 0\,0\,0 & 0\,0\,0 & \\ & [A_1] & 0\,0\,0 & 0\,0\,0 & \\ & & [A_1] & 0\,0\,0 & \cdots \\ & & & [A_1] & \\ & & & & \end{bmatrix} \tag{6.3a}$$

or

$$G = \begin{bmatrix} 1\,1\,1 & 0\,1\,1 & 0\,0\,1 & 0\,0\,0 & 0\,0\,0 & \\ 0\,0\,0 & 1\,1\,1 & 0\,1\,1 & 0\,0\,1 & 0\,0\,0 & \\ 0\,0\,0 & 0\,0\,0 & 1\,1\,1 & 0\,1\,1 & 0\,0\,1 & \ldots \\ & & & 1\,1\,1 & 0\,1\,1 & \\ & & & & 1\,1\,1 & \end{bmatrix} \qquad (6.3b)$$

where the blank areas indicate all-zeros entries, and the matrix G extends arbitrarily far down and to the right. In practice, of course, G would be finite but, in a meaningful case, very large. This matrix G is a generator matrix in exactly the same sense as for block code. That is, any code (i.e., transmitted) sequence can be represented as a linear combination of the rows of G.

Now look carefully at the rows of G. Notice that they are simply shifted replicas of one another. Because each row by itself is certainly a code sequence, you should be able to find that sequence in the tree for this code. Inspection of Figure 6.3 shows this to be the case. The same argument applies to any linear combination of two or more rows of G. Thus, given the generator matrix, you can generate the tree for the code.

The process can also be reversed because a one-to-one correspondence exists between G and the tree. This gives a second way of obtaining G from other information about the code. As you follow along one or more successive all-zeros branches and finally choose the first branch corresponding to a data bit equal to 1, note that this latest branch is always 111, regardless of where you are in the tree (including, you may notice, the branch of 1s at the initial node). Thus, the path to node A′ is

111 011 001

which corresponds to the data bit sequence 100, while data bit sequence 0100 gives the path

000 111 011 001,

which is simply the sequence to node A′ shifted three digits to the right. Further shifts of this type occur if you take data sequences 00100, 000100, *et cetera.* This suggests that such shifted versions of the same sequence might be used as rows of G, provided that all sequences in the tree can be produced from some smaller number of row vectors. To determine whether this can actually be done, try using as a basis the following vector **a** and its translations **b** and **c:**

$$\mathbf{a} = 111\ 011\ 001\ 000\ \ldots$$

$$\mathbf{b} = 000\ 111\ 011\ 001\ 000\ \ldots$$

$$\mathbf{c} = 000\ 000\ 111\ 011\ 001\ 000\ \ldots$$

where all additional components are 0s. Now let the path to node A in Figure 6.3 be labeled #1, that to node B, #2, *et cetera,* with #8 assigned to the path to node D′. Comparing vectors **a, b,** and **c,** truncated after $n_0K = 9$ digits, with these paths, observe that $\mathbf{a} = \#5$, $\mathbf{b} = \#3$, and $\mathbf{c} = \#2$. With this bit of encouragement, check to see whether linear combinations of **a, b,** and **c** give rise to the remaining paths in the tree:

$$\begin{aligned}
\mathbf{a} + \mathbf{b} &= 111\ 100\ 010\ 001\ 000\ \ldots \\
\mathbf{a} + \mathbf{c} &= 111\ 011\ 110\ 011\ 001\ 000\ \ldots \\
\mathbf{b} + \mathbf{c} &= 000\ 111\ 100\ 010\ 001\ 000\ \ldots \\
\mathbf{a} + \mathbf{b} + \mathbf{c} &= 111\ 100\ 101\ 010\ 001\ 000\ \ldots
\end{aligned}$$

Again comparing the nine-digit truncated vectors with the tree, observe that these last four sums have generated paths 7, 6, 4, and 8, respectively.

To conclude this section, consider generalizing from $R = 1/n_0$ to $R = k_0/n_0$ with $k_0 \geqslant 2$. To do this, let there be k_0 parallel inputs to the encoder, and denote the transforms of these input sequences by $I^{(1)}(D)$, $I^{(2)}(D), \ldots, I^{(k_0)}(D)$. For a nonsystematic code, each transmitted sequence will be a linear combination of all k_0 input sequences and will have the following transform:

$$\begin{aligned}
T^{(j)}(D) = {} & G^{(j,1)}(D)\, I^{(1)}(D) + G^{(j,2)}(D) I^{(2)}(D) + \ldots \\
& + G^{(j,k_0)}(D) I^{(k_0)}(D)
\end{aligned}$$

for $j = 1, 2, \ldots, n_0$. Notice that there can be up to k_0n_0 generator polynomials $G^{(j,\ell)}(D)$. Each polynomial specifies the tap connections from one shift register to one output (transmitted) sequence line. An encoder for a nonsystematic convolutional code of rate k_0/n_0 is shown in block diagram form in Figure 6.6. Each SR is of length K and has n_0 sets of taps, one set for each transmitted sequence. For a systematic code, there are only $k_0(n_0 - k_0)$ nontrivial generator polynomials because $T^{(j)}(D) = I^{(j)}(D)$ for $j = 1, 2, \ldots, k_0$.

The matrix representation of encoding becomes somewhat more cumbersome in detail. The input message vector will be written as

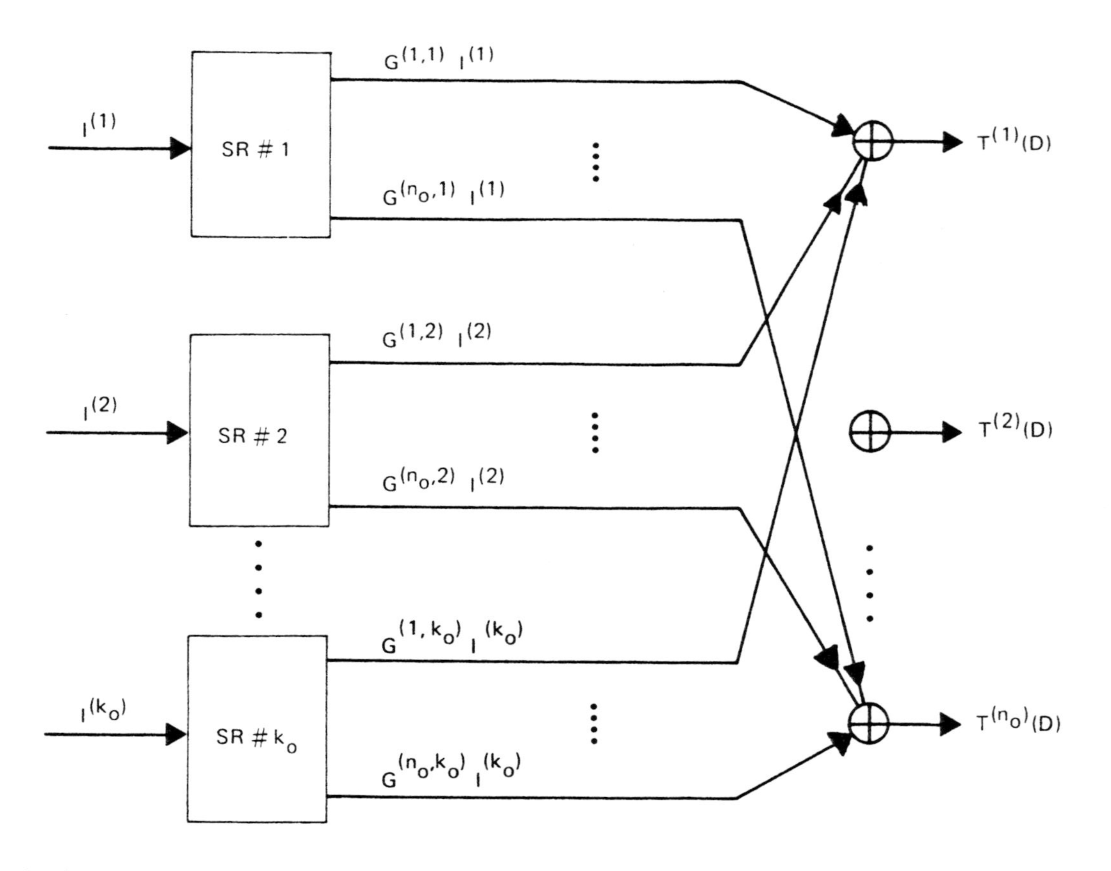

Figure 6.6 Encoder block diagram for $k_0 > 1$.

$$\mathbf{u} = [u_1^{(1)}u_1^{(2)} \ldots u_1^{(k_0)}u_2^{(1)}u_2^{(2)} \ldots u_2^{(k_0)} \ldots] \tag{6.4}$$

where the subscript refers to time, as before, and the superscript identifies the input sequence. Similarly, the output vector will be

$$\mathbf{x} = [x_1^{(1)}x_1^{(2)} \ldots x_1^{(n_0)}x_2^{(1)}x_2^{(2)} \ldots x_2^{(n_0)} \ldots] \tag{6.5}$$

The structure of the generator matrix G is also modified, of course. It can now be written as

$$G = \begin{bmatrix} G_1 & G_2 & G_3 \ldots & & & G_K \ldots & & \\ & G_1 & G_2 & G_3 \ldots & & & G_K & \ldots \\ & & G_1 & G_2 & G_3 & \ldots & & G_K \quad \ldots \\ & & \cdot & & & & \cdot & \\ & & & \cdot & & & & \cdot \\ & & & & \cdot & & & \end{bmatrix} \tag{6.6}$$

Here each submatrix G_i contains k_0 rows and n_0 columns. All other entries in G are zero. An example using a generator matrix for $k_0 > 1$ will be given in Section 6.6.

6.4 DISTANCE PROPERTIES

Recall that for block codes, the concept of distance between two vectors provides a way of quantitatively describing the extent to which the code words and block-length sequences are different. For block codes, the most commonly used distance function is Hamming distance—the number of positions in which two words (or block-length sequences) differ. It was shown that a code should have the largest possible minimum distance between code words.

Convolutional codes also have a distance concept defined for them. In fact, two types of distance are in use, each of which is important. The distance between words of block length n corresponds to the distance between sequences of various lengths that differ in their initial information bit i_0, again given as Hamming distance. The first type of distance defined in this way is the distance between encoded sequences of length n_0K, corresponding to the code constraint length K. The smallest value of this distance, denoted $d_{\min}$, has always been an important code parameter. (For many years, this was the only distance concept used for convolutional codes.)

Massey [6.2] coined the term "free distance," d_{free}, to denote a type of distance that was found to be an important parameter for two of the decoding techniques used for convolutional codes. *Free distance* is defined to be the minimum distance between arbitrarily long (i.e., possibly infinite) encoded sequences, one corresponding to an initial information bit $i_0 =$ 0 and one to $i_0 = 1$. The definitions of d_{free} and d_{min} in terms of code sequences corresponding to $i_0 = 0$ and $i_0 = 1$ are of interest because one of these values of i_0 represents correct decoding, while the other represents incorrect decoding. Because of linearity, you can choose $i_0 = 0$ as the correct value without loss of generality. Going one step further, define the *correct subset* S_C and *incorrect subset* S_I as the sets of all encoded sequences corresponding to $i_0 = 0$ and $i_0 = 1$, respectively. Also because of linearity, the encoded sequence corresponding to the difference of a sequence in S_C and a sequence in S_I is itself in S_I because it corresponds to an information sequence beginning with a 1. Thus, in determining d_{min} and d_{free}, we need only consider sequences in S_I. Therefore, d_{min} and d_{free} are the minimum number of 1s in encoded sequences of lengths n_0K and infinity, respectively, which are members of S_I.

Alternative definitions of free distance can be given in terms of the state diagram and trellis. While all three definitions are equivalent, the latter two may often be less cumbersome from a computational point of view. In terms of the state diagram, d_{free} is the total weight of the output sequence corresponding to a series of transitions that begin and end with the all-0 state and include at least one other state. The corresponding definition based on the trellis reveals that d_{free} is the minimum weight of the output sequence on a path that leaves the all-0 state and merges again with the all-0 sequence at some point further out in the trellis.

Because free distance is defined in terms of an infinitely long encoded sequence, it particularly influences convolutional code performance with a decoding algorithm that can use searches over more than one constraint length—i.e., sequential and Viterbi decoding. Obviously, it is always true that $d_{min} \leq d_{free}$; it is surprising that $d_{min} = d_{free}$ for many codes.

Example 6.6. To illustrate the notions of minimum distance and free distance, refer again to the systematic code defined by the tree in Figure 6.3. To determine d_{min}, it is necessary only to find the weights for the four code sequences corresponding to information sequences of length $K = 3$, which terminate at nodes A′, B′, C′, and D′. Let w(A′), *et cetera*, denote the weights of these code sequences. It is easy to see that w(A′) = 6, w(B′) = 7, w(C′) = 5, and w(D′) = 6. Thus, $d_{min} = 5$. Now, can we find d_{free} without too much difficulty? In this case, yes. Observe that the weight of a code sequence stops growing once that sequence corresponds to an all-zeros information sequence. This is already the case at A′ if the upper branch is chosen each time. If $w(A', \infty)$, *et cetera*, denotes the minimum weight of any infinitely long code sequence starting at the specified node (e.g., A′), then $w_{min}(A', \infty) = 0$, while $w_{min}(B', \infty)$, $w_{min}(C', \infty)$, and $w_{min}(D', \infty)$ are all at least equal to 1. Adding each of these weights to the weight of the code sequence from the base of the tree out to the corresponding node shows that the minimum-weight infinite-length sequence through A′ has $w = 6$, while for all other paths $w \geqslant 6$. Thus $d_{free} = 6$. You can check this result with Figure 6.4 or Figure 6.5 for paths abca and abdca.

Searching for a code having the largest minimum distance is at best tedious and at worst practically impossible by hand; even with the aid of a computer it can become quite time-consuming due to the exponential growth (with K) in the number of sequences to be examined. Fortunately, a number of clever ways can reduce this effort, particularly by reducing the number of codes that need to be examined in order to determine the optimum (i.e., maximum d_{min} or d_{free}) code for a given constraint length K. A number of computer-based search techniques have been developed. Results of applying these techniques are displayed in the tables of Section 6.6. Especially useful, of course, are those cases in which the code generators are given along with values of the distance parameters.

6.5 DISTANCE BOUNDS FOR CONVOLUTIONAL CODES

Both upper and lower bounds can be computed on the minimum distance of a convolutional code having rate $R = k_0/n_0$ and constraint length $= n_0K$. These bounds are similar in nature and derivation to those for block codes, with block length corresponding to constraint length. Unfortunately, in general, neither the upper nor lower bound is very tight, so that the bounds' usefulness is generally restricted to giving a rough idea of how good a code is in terms of how closely its minimum distance approaches the values of minimum distance given by the bounds. The bounds are presented without proof, and only for the case of binary codes.

For the lower bound on d, there is a Gilbert-like bound characterized as follows. For rate R and constraint length n_0K, let d be the largest integer that satisfies

$$H\left(\frac{d}{n_0K}\right) \leqslant 1 - R \tag{6.7}$$

Then at least one binary convolutional code exists with minimum distance d for which (6.7) holds. Here, $H(x)$ is the familiar entropy function for a binary alphabet:

$$H(x) = -x \log_2 x - (1 - x) \log_2 (1 - x), \quad 0 \leq x \leq 1$$

One upper bound on minimum distance corresponds to the Plotkin bound for block codes. It can be stated as follows. For a binary code with $R = 1/n_0$, the minimum distance d_{min} satisfies

$$d_{min} \leq \lfloor (1/2)(n_0 K + n_0) \rfloor \tag{6.8}$$

where $\lfloor I \rfloor$ denotes the largest integer equal to or less than I—that is, the integer portion of I.

Heller [6.3] found a simple upper bound on d_{free}. It is given by

$$d_{free} \leq \min_{j \geq 1} \left\lfloor \frac{n_0}{2} \frac{2^j}{2^j - 1} (K + j - 1) \right\rfloor \tag{6.9}$$

Consider the calculation of this bound on d_{free}. For any given case, n_0 and K are fixed because they are parameters of the code. The fraction $2^j/(2^j - 1)$ quickly approaches the value 1 from above with increasing j. The expression $(K + j - 1)$ obviously grows linearly with j. Thus, the minimum value of the expression in brackets should be reached for a small value of j. This can be verified analytically by regarding d_{free} and j in (6.9) as continuous variables and applying the usual first-derivative test.

As stated earlier, these bounds are usually quite coarse, although occasionally you can find a code that actually achieves one of them.

Example 6.7. Apply the bounds of (6.7), (6.8), and (6.9) to the code of Figure 6.3 and Example 6.6. $n_0 = 3$, $K = 3$, $R = 1/3$, with $d_{min} = 5$ and $d_{free} = 6$. You can thus obtain the following information:

a. The Gilbert bound becomes
$H(d/9) \leq 2/3$

From direct calculation or look-up table (e.g., Fano [6.4], Appendix B), you find out that

$$H(0.1735) = 0.6656$$
$$H(0.1740) = 0.6668$$

so that $d/9 \leq 0.1740$, $d \leq 1.566$. Thus, the largest integer d that satisfies (6.4) is $d = 1$, giving $d_{min} \geq 1$, obviously a poor lower bound in this case.

b. After substitution, the Plotkin bound expression becomes

$$d_{min} \leq \lfloor (1/2)(9 + 3) \rfloor = 6$$

which is a good upper bound.

c. The Heller bound on d_{free} is tabulated below for $j = 1, \ldots, 5$. From these values you see that $d_{\text{free}} \leq 8$, again a rather good bound.

j	$\lfloor \cdot \rfloor$	
1	$\lfloor 9 \rfloor$	$= 9$
2	$\lfloor 8 \rfloor$	$= 8$
3	$\lfloor 60/7 \rfloor$	$= 8$
4	$\lfloor 48/5 \rfloor$	$= 9$
5	$\lfloor 336/31 \rfloor$	$= 10$

6.6 KNOWN GOOD CONVOLUTIONAL CODES

This section presents tables of optimum or nearly optimum codes as they have appeared in the literature. In some cases, the author's original notation has been modified so that it conforms with our own. You will notice the terms *catastrophic* and *noncatastrophic code* used in conjunction with the codes by Larsen in Tables 6.2 through 6.4. This terminology will be explained in Chapter 7. Lack of familiarity with the concept will not present any difficulty for the present discussion.

Larsen [6.5] tabulated noncatastrophic codes having rates $R = 1/2$, $1/3$, $1/4$ and $3 \leq K \leq 14$, and catastrophic codes for $R = 1/2$, $K = 5, 12$, and 14. These codes are given in Tables 6.2, 6.3, and 6.4. Some of the codes listed had been published earlier by Odenwalder [6.6] while the majority of them were more recent results by Larsen. The bound referred to is equal to or 1 greater than the Heller bound discussed in Section 6.5. In cases where this bound is not actually achieved, an exhaustive search of all codes established optimality. The generator is given as the sequence $1, g_1^{(i)}, g_2^{(i)}$, *et cetera*, where

$$G^{(i)}(D) = 1 + g_1^{(i)}D + g_2^{(i)}D^2 + \ldots + g_{K-1}^{(i)}D^{K-1}$$

is the transform of the code generator. Note that all of the codes listed are nonsystematic. In Tables 6.2, 6.3, and 6.4, $\nu = K$ and $N = n_0 K$ as used here.

Example 6.8. The rate 1/2 noncatastrophic code having $K = 4$ has $n_A = K \times 1/R = 8$ and has generators given in octal notation by 15 and 17, corresponding respectively to $G^{(1)}(D) = 1 + D + D^3$ and $G^{(2)}(D) = 1 + D + D^2 + D^3$. This code has $d_{\text{free}} = 6$, which achieves the bound given by (6.6) for these code parameters. (The reader should verify this last statement.)

Table 6.2 Rate 1/2 Codes with Maximum Free Distance

A. Noncatastrophic Codes

ν	N	generators(octal)		d_{free}	bound
3	6	5	7[1]	5	5
4	8	15	17[1]	6	6
5	10	23	35[1]	7	8
6	12	53	75[1]	8	8
7	14	133	171[1]	10	10
8	16	247	371[1]	10	11
9	18	561	753[1]	12	12
10	20	1167	1545	12	13
11	22	2335	3661	14	14
12	24	4335	5723	15	16
13	26	10533	17661	16	16
14	28	21675	27123	16	17

B. Catastrophic Codes

ν	N	generators(octal)		d_{free}	bound
5	10	27	35	8	8
12	24	5237	6731	16	16
14	28	21645	37133	17	17

Reproduced with permission from Larsen [6.5]

[1]This code was found by Odenwalder [6.6] and is listed here for completeness.

Table 6.3 Rate 1/3 Noncatastrophic Codes with Maximum Free Distance

	N	generators (octal)			d_{free}	bound
3	9	5	7	7[1]	8	8
4	12	13	15	17[1]	10	10
5	15	25	33	37[1]	12	12
6	18	47	53	75[1]	13	13
7	21	133	145	175[2]	15	15
8	24	225	331	367[1]	16	16
9	27	557	663	711	18	18
10	30	1117	1365	1633	20	20
11	33	2353	2671	3175	22	22
12	36	4767	5723	6265	24	24
13	39	10533	10675	17661	24	24
14	42	21645	35661	37133	26	26

[1]This code was found by Odenwalder [6.6] and is listed here for completeness.
[2]This code was also found by Odenwalder [6.6], but was overlooked. The corresponding code in [6.6] has free distance only 14.

Table 6.4 Rate 1/4 Noncatastrophic Codes with Maximum Free Distance

ν	N	generators(octal)				d_{free}	bound
3	12	5	7	7	7	10	10
4	16	13	15	15	17	13	13
5	20	25	27	33	37	16	16
6	24	53	67	71	75	18	18
7	28	135	135	147	163	20	20
8	32	235	275	313	357	22	22
9	36	463	535	733	745	24	24
10	40	1117	1365	1633	1653	27	27
11	44	2387	2353	2671	3175	29	29
12	48	4767	5723	6265	7455	32	32
13	52	11145	12477	15573	16727	33	33
14	56	21113	23175	35527	35537	36	36

Reproduced with permission from Larsen [6.5].

Earlier, Bahl and Jelinek [6.7] published two tables of rate 1/2 systematic convolutional codes the free distance of which is optimum for systematic codes and is equal to the free distance for optimum known nonsystematic codes at $\nu = K = 3, 4, 5$, and 6. Data from these two tables are reproduced here in Tables 6.5 and 6.6. Table 6.5 gives all codes found for $3 \leqslant K \leqslant 9$, while Table 6.6 presents only one code for each K (where $\nu = K$). This latter table resulted from taking, in those cases when more than one code was found having the maximum d_{free}, that code with the largest $d_{\min}$ the generator of which had the smallest weight W (the last column in Table 6.6). The purpose of this last criterion was the minimization of the number of adders in the encoding circuitry.

A complementary code of rate ½ (as in Table 6.6) is defined as follows. The high-order and low-order coefficients $g_{K-1}^{(i)}$ and $g_0^{(i)}$ are all 1, and the remaining coefficients of one generator polynomial are the binary complements of those in the other generator polynomial. For example, the first code of Table 6.6 has generator polynomials $1 + D^2$ and $1 + D + D^2$ corresponding to 101 and 111.

Tables 6.7 through 6.9 present results obtained by Paaske [6.8] in a computer-based search for optimal (in the sense of having maximum d_{free}) short convolutional codes of rates 2/3 and 3/4. Paaske used several criteria based on theoretical results obtained in his paper to discard unpromising codes quickly, thus reducing the search effort. Note that some of the upper bounds on free distance in Table 6.7 were, in fact, not achieved. The parameter ν in Tables 6.7, 6.8, and 6.9 is equal to our $K - 1$.

Table 6.5 Generators of Codes with $d_f = \nu + 2$

ν	Generator
3	101
4	1011
5	10011
6	100011
	100101
7	1000101
	1001101
	1010011
8	10001001
	10011011
	10011101
	10101101
9	100010011
	100011011
	100100011
	100101001
	100110101
	101000011
	101001011

Reproduced with permission from Bahl and Jelinek [6.7].

Table 6.6 $R = 1/2$ Complementary Codes

ν	Generator (Octal)	d_{free}	d_{min}	Weight
3	5	5	3	2
4	13	6	3	3
5	31	7	4	3
6	61	8	4	3
7	121	9	5	3
8	211	10	5	4
9	503	11	6	4
10	1065	12	6	5
11	2415	13	7	5
12	5121	14	7	5
13	12043	15	7	5
14	24421	16	8	5
15	51303	17	7	7
16	120643	18	8	7
17	346411	18	9	8
18	425551	20	8	9
19	1411041	20	9	6
20	2734605	20	10	11
21	5011303	22	9	8
22	11047441	22	10	9
23	22517023	24	10	11
24	51202215	24	10	9

Reproduced with permission from Bahl and Jelinek [6.7].

Table 6.7 Upper Bounds for Short Rates 2/3 and 3/4 Convolutional Codes

constraint length	free distance			
	R=2/3		R=3/4	
ν	upper bound	achieved	upper bound	achieved
2	4	3		
3	4	4	4	4
4	6	5	4	(4)
5	6	6	6	5
6	8	7	6	6
7	8	8	8	(6)
8	8	(8)	8	7
9	10	9	8	8
10	10	10	9	

Reproduced with permission from Paaske [6.8].

Table 6.8 Generator Matrices of the (3,2) Convolutional Codes in Table 6.7

ν	$\underline{G}_0$	$\underline{G}_1$	$\underline{G}_2$	$\underline{G}_3$	$\underline{G}_4$	$\underline{G}_5$
2	101 011	111 100				
3	101 011	011 001	000 101			
4	101 011	100 101	110 011			
5	101 011	111 001	011 101	000 101		
6	101 011	111 111	010 101	101 011		
7	101 011	110 001	011 101	011 111	000 110	
9	101 011	001 010	101 011	011 100	110 001	000 101
10	101 011	100 111	010 100	011 010	101 100	110 011

Reproduced with permission from Paaske [6.8].

Table 6.9 Generator Matrices of the (4,3) Convolutional Codes in Table 6.7

ν	$\underline{G}_0$	$\underline{G}_1$	$\underline{G}_2$	$\underline{G}_3$
3	1111 0101 0011	0000 0110 0100	0000 0000 0011	
5	1001 0101 0011	1111 0101 0100	0000 1001 0011	
6	1001 0101 0011	1001 1001 1110	0101 1010 0110	
8	1001 0101 0011	1110 0000 0010	1100 1101 0110	0000 1001 1010
9	1001 0101 0011	0011 0111 1011	0110 0001 1000	0110 1100 1001

Reproduced with permission from Paaske [6.8].

Johannesson [6.9] discovered both systematic and nonsystematic optimum codes of short to medium length, all at rate 1/2. He defined several types of optimality. A code is called an *optimum minimum distance* (OMD) code when its minimum distance $d_{\min}$ is equal to or greater than that of any code of the same constraint length. Similarly, an *optimum free distance* (OFD) code has its free distance d_{free} equal to or greater than that of any code of the same constraint length. To define the third type of optimality, the concept of distance profile must first be defined. The distance profile of a convolutional code is the set of minimum weights of paths resulting from an information sequence having $i_0 = 1$, taken over path lengths (measured in branches) of 1, 2, . . . , K. Then a code is said to be an *optimum distance profile* (ODP) code if its distance profile is equal or superior to the distance profile of any code of the same constraint length.

A *quick-look-in* (QLI) code is a nonsystematic code with information digits that can be obtained simply by adding the two encoded sequences together modulo 2. Thus, without actually decoding, it is possible to get an idea of the transmitted information bits. This capability may be highly desirable in a situation in which you need at least a rough idea of the data being sent, at the earliest possible time. Obviously, you have no way of knowing whether errors have been committed when using the quick-look-in technique, but if the E_B/N_0 is known for the channel, you at least know and can usually accept the average bit error rate.

Some explanation of the notation used in Tables 6.10 through 6.14 must be given. Johannesson's M is our $K - 1$, and his d_M and d_∞ are our d_{min} and d_{free}, respectively. His generators $G^{(1)}$ and $G^{(2)}$ are given in octal notation with the highest-order bit corresponding to the lowest-degree term in the generator polynomial. Furthermore, octal representation of generators are all left justified (i.e., the high-order bit positions of the left-most octal digit are always filled in), which is just the reverse of the notation commonly used. The notation "# paths" refers to the number of paths in the tree that have the distance shown.

Table 6.10 ODP Systematic Convolutional Codes with Rate 1/2, which are also OMD Codes

M	$G^{(2)}$		d_M	#paths	d_∞	#paths
1	6	B	3	2	3	1
2	7	B	3	1	4	2
3	64	B	4	3	4	1
4	72	B	4	1	5	2
5	73	B	5	5	6	3
6	730	B	5	2	6	3
	734		5	3	6	1
7	724	B	6	11	6	2
8	715	B	6	5	7	2
	671		6	6	7	1
9	6710	B	6	1	7	1
	7154		6	3	8	4
10	6710	B	7	12	7	1
	7152		7	13	8	3
11	6711	B	7	5	8	2
	7153		7	6	9	3
12	67114	B	8	29	9	1
13	67114	B	8	12	9	1
14	67115		8	6	10	4

Reproduced with permission from Johannesson [6.9].

A look at the two generators for any nonsystematic QLI code reveals that they are identical except for the single bit that is second from the left; this position, which corresponds to D in the generator polynomial, is present in $G^{(1)}$ but not in $G^{(2)}$. Thus, $G^{(1)} - G^{(2)} = D$, always, so the QLI sequence is

$$
\begin{aligned}
&(G^{(1)}(D) + G^{(2)}(D))\, I(D) + E^{(1)}(D) + E^{(2)}(D) \\
&= D \cdot I(D) + E^{(1)}(D) + E^{(2)}(D)
\end{aligned}
$$

which gives the information sequence one time unit after the two received sequences enter the decoder's encoder, corrupted by the mod-2 sum of the two noise sequences $E^{(1)}(D)$ and $E^{(2)}(D)$.

Table 6.11 ODP Systematic Convolutional Codes with Rate 1/2

M	$G^{(2)}$	d_M	#paths	d_∞	#paths
15	714474	8	1	10	1
16	714476	9	18	10	1
	671166	9	22	12	13
17	671145	9	7	11	1
	671166	9	13	12	13
18	6711454	9	3	12	4
19	7144616	10	31	12	3
20	7144616	10	13	12	3
	7144761	10	18	12	1
21	67114544	10	4	12	1
22	71446162	10	1	13	2
	71446166	10	6	14	6
23	67114543	11	27	14	6
	67115143	11	32	14	2
24	714461654	11	11	15	5
	671151434	11	16	15	4
25	714461654	11	5	15	5
	671145536	11	9	15	3
26	671145431	11	1	15	1
	671151433	11	4	16	8
27	7144616264	12	21	14^L	1
	7144760524	12	26	16	7
28	6711454306	12	8	16	4
	6711514332	12	13	16	3
29	7144616573	12	2	17^L	3
	7144760535	12	6	18	22
30	71446162654	13	43	16^L	2
	67114543064	13	44	16^L	1
31	71446162654	13	15	16L	2
	67114543066	13	24	18	11
32	71446162655	13	4	17^L	2
	71447605247	13	13	18^L	2
33	714461626554	13	1	18^L	5
	671145430654	13	4	18^L	1
34	714461626554	14	34	18^L	5
	714461625306	14	42	18^L	1
35	714461625313	14	14	18^L	3
	714461626555	14	19	19^L	2

Note: L denotes that this number is actually d_{71} which is a lower bound on d_∞.

Reproduced with permission from Johannesson [6.9].

Table 6.12 ODP QLI Codes with Rate 1/2

M	$G^{(1)}$	$G^{(2)}$	d_M	#paths	d_∞	#paths
1	6	4	3	2	3	1
2	7	5	3	1	5	1
3	74	54	4	3	6	1
4	72	52	4	1	6	1
5	71	51	5	5	7	1
	75	55	5	6	8	2
6	704	504	5	2	7	1
	714	514	5	3	8	1
7	742	542	6	11	9	1
8	742	542	6	5	9	1
9	7404	5404	6	1	9	1
	7434	5434	6	2	10	2
10	7406	5406	7	12	10	1
	7422	5422	7	13	11	2
11	7421	5421	7	5	11	1
	7435	5435	7	6	12	5
12	74044	54044	8	29	11	1
13	74042	54042	8	12	11	1
	74046	54046	8	17	13	2
14	74042	54042	8	6	11	1
	74047	54047	8	7	14	2
15	740414	540414	8	1	13	1
	740470	540470	8	3	14	2
16	740416	540416	9	18	14	1
	740462	540462	9	22	15	3
17	740415	540415	9	7	15	3
	740463	540463	9	9	16	2
18	7404244	5404244	9	3	15	1
	7404634	5404634	9	4	16^L	1
19	7404242	5404242	10	31	15	1
20	7404241	5404241	10	13	14^L	1
	7404155	5404155	10	18	18^L	2
21	74042404	54042404	10	4	15	1
	74041550	54041550	10	8	18^L	2
22	74041566	54041566	10	1	18	1
	74042436	54042436	10	8	19^L	2
23	74042417	54042417	11	27	18^L	1
	74041567	54041567	11	32	19^L	1

Note: L denotes that this number is actually d_{71} which is a lower bound on d_∞.

Table 6.13 Nonsystematic QLI Codes with Maximum Free Distance for QLI Codes

M	$G^{(1)}$	$G^{(2)}$		d_M	#paths	d_∞	#paths
1	6	4	1,2,3	3	2	3	1
2	7	5	1,2,3	3	1	5	1
3	74	54	1,2,3	4	3	6	1
4	66	46	1, 3	4	2	7	2
5	75	55	1,2,3	5	6	8	2
6	654	454	3	5	3	9	4
7	742	542	2,3	6	11	9	1
8	751	551	3	6	7	10	1
9	7664	5664		5	1	11	3
10	7506	5506	3	7	14	12	3
11	7503	5503		6	2	13	8
12	76414	56414		7	7	14	10
13	66716	46716		6	1	14	3

Notes: 1. This code is OFD.
2. This code is ODP.
3. This code is OMD.

Reproduced with permission from Johannesson [6.9]

Table 6.14 Nonsystematic Codes that are simultaneously ODP, OMD, and OFD

M	$G^{(1)}$	$G^{(2)}$	d_M	#paths	d_∞	#paths
1	6	4	3	2	3	1
2	7	5	3	1	5	1
3	74	54	4	3	6	1
4	62	56	4	2	7	2
5	75	55	5	6	8	2
6	634	564	5	3	10	12
7	626	572	6	11	10	1
8	751	557	6	6	12	10
9	7664	5714	6	2	12	1
10	7512	5562	7	13	14	19
11	-	-	7	-	15	-
12	-	-	8	-	16	-
13	60676	45662*	8	17	16	5

* The search for the code with the smallest number of d_∞=16 paths was not exhaustive and hence a slightly better code might exist.

Reproduced with permission from Johannesson [6.9].

Example 6.9. Consider the code specified by the third line of Table 6.14. This code has $K = 4$ with the following encoder tap connections:

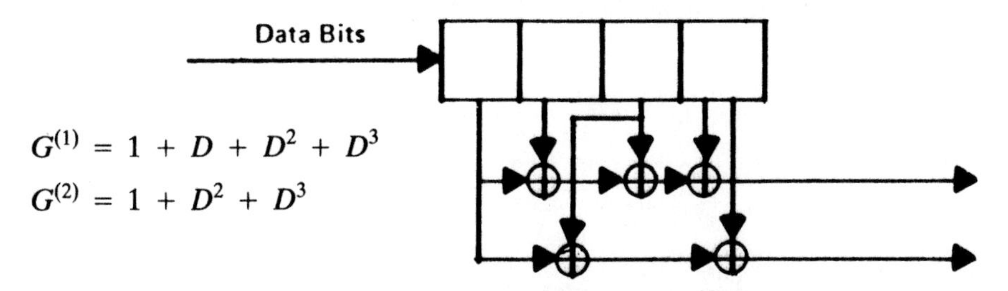

$$G^{(1)} = 1 + D + D^2 + D^3$$
$$G^{(2)} = 1 + D^2 + D^3$$

Johannesson [6.9] has also compared minimum and free distances for his various classes of optimum codes with these parameters for six classes of codes found earlier by others (as indicated), and with the Gilbert lower bound. Figures 6.7(a) and 6.7(b) together present all of these comparisons.

Other tables of optimum convolutional codes are given in Lin and Costello [5.8], Chapter 11, and in Clark and Cain [6.10], Appendix B.

Example 6.10. As a final example in this chapter, let us derive the generator matrix and encoder for the rate 2/3 code having $K = 2$ in Table 6.8. Here

$$G_1 = \begin{bmatrix} 1 & 0 & 1 \\ 0 & 1 & 1 \end{bmatrix} \qquad G_2 = \begin{bmatrix} 1 & 1 & 1 \\ 1 & 0 & 0 \end{bmatrix}$$

so that

$$G = \begin{bmatrix} 1 & 0 & 1 & 1 & 1 & 1 & & & & \\ 0 & 1 & 1 & 1 & 0 & 0 & & & & \\ & & & 1 & 0 & 1 & 1 & 1 & 1 & \\ & & & 0 & 1 & 1 & 1 & 0 & 0 & \cdot \\ & & & & & & 1 & 0 & 1 & \cdot \\ & & & & & & 0 & 1 & 1 & \cdot \end{bmatrix}$$

Each block of $n_0 = 3$ columns specifies encoder ouput at some instant of time. Alternate rows (for $k_0 = 2$) specify the stages of one shift register that are to be added modulo 2 into the output line corresponding to that column. The shift register tap connections can be deduced by noting that, for any column in a block of k_0 columns, odd and even numbered rows correspond to SR #1 and SR #2, respectively. The result is shown in Figure 6.8.

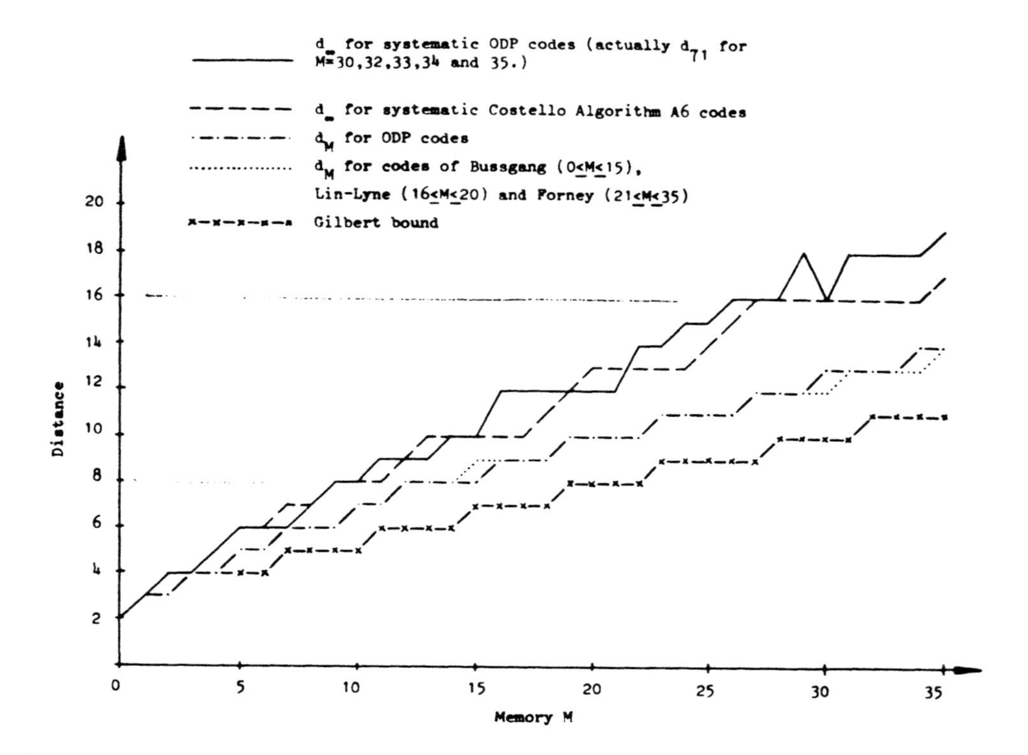

Figure 6.7(a) Minimum distance d_M and free distance d_∞ for some rate 1/2 convolutional codes. Reproduced with permission from Johannesson [6.9].

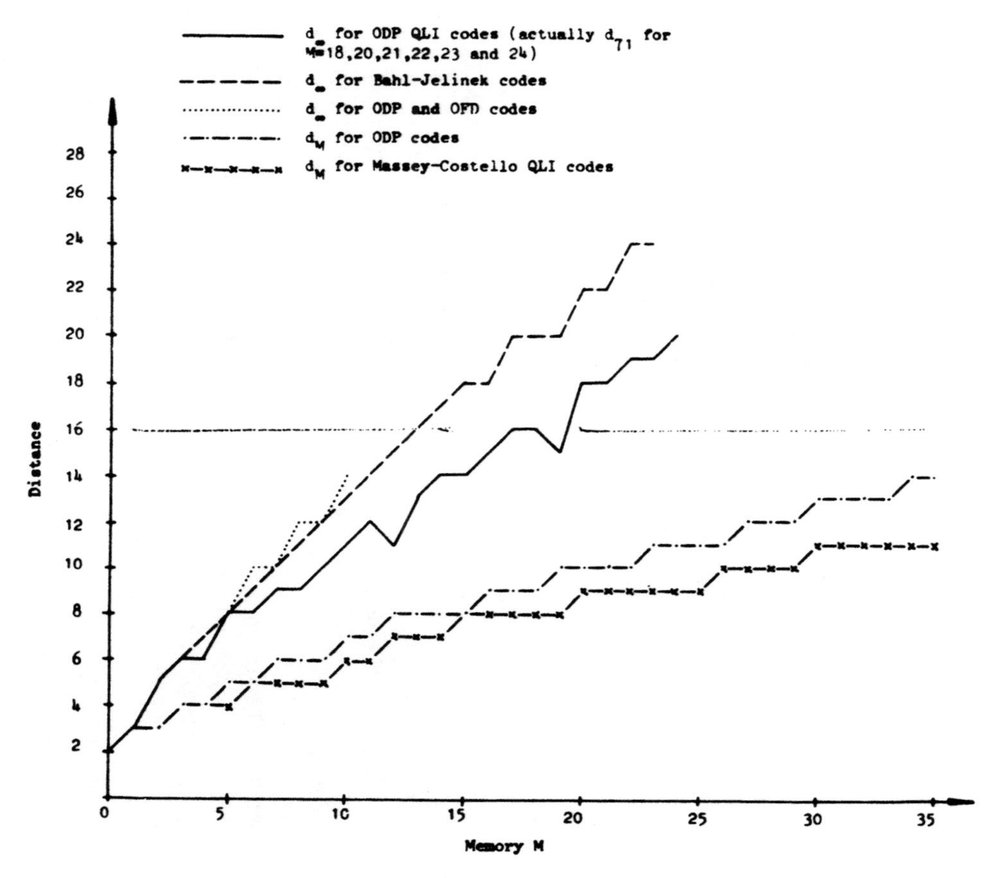

Figure 6.7(b) Minimum distance d_M and free distance d_∞ for some rate 1/2 convolutional codes. Reproduced with permission from Johannesson [6.9].

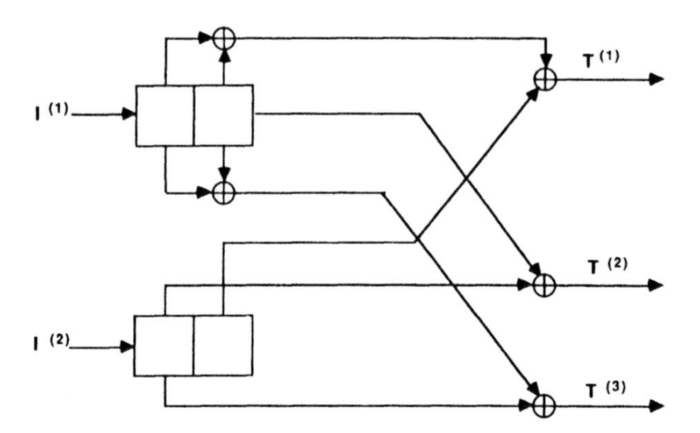

Figure 6.8 Encoder for $R = 2/3$, $K = 2$ binary convolutional code.

6.7 NONBINARY CONVOLUTIONAL CODES

Dual-m codes are nonbinary convolutional codes using an alphabet the symbols of which are elements of GF(2^m) and encoding with a two-stage SR in which each stage holds a 2^m-ary symbol. Thus, for example, a dual-3 code groups three parallel binary inputs into an 8-ary symbol, and the encoder consists of two 8-ary stages and appropriate taps. Dual-m codes are especially well matched to the use of 2^m-ary modulation because each output symbol directly specifies which of 2^m waveforms should be transmitted. An encoder for an optimum, rate ½, dual-3 code is shown in Figure 6.9. Note the numbering scheme for the three lines of each output symbol. The specification of tap connections can still be done compactly, just as for binary codes, but a bit more explanation of the notation is needed. For instance, Clark and Cain [6.10] in their Table B-6 give the tap connections for the encoder in Figure 6.9 as follows:

(11, 22, 44)

(41, 24, 13)

Each set of three octal numbers in parentheses specifies connections for one output symbol. Within a set of parentheses, each two-digit octal number indicates connections to a line (or bit) of the corresponding output symbol (1, 2, or 3). The first digit of a pair of octal digits specifies output from the first stage or symbol of the encoder SR, and the second digit specifies the output from the second stage. In each octal digit, the positions of the binary 1s (third or left, second or middle, first or right) specify the

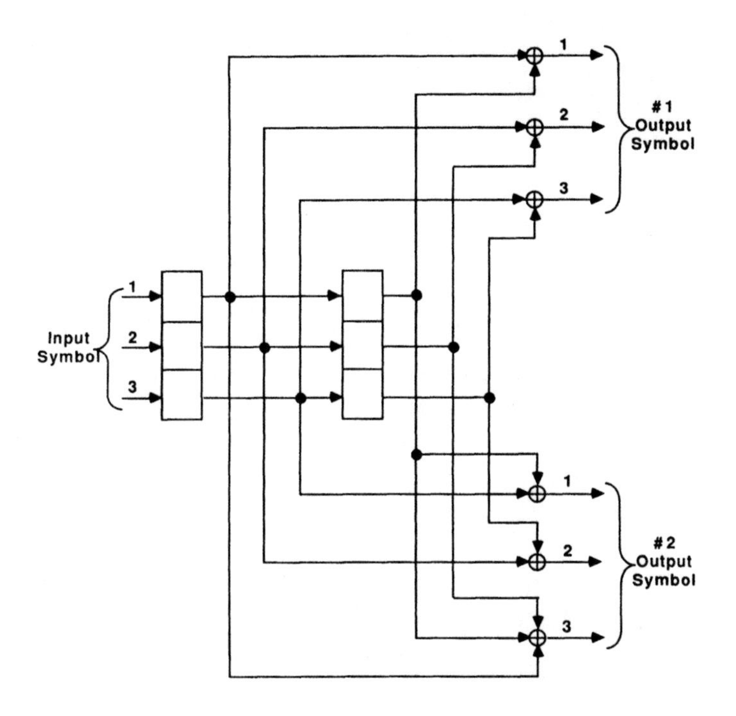

Figure 6.9 Encoder for $R = ½$, dual-3 convolutional code.

bit positions within the corresponding 8-ary SR stage, which are summed modulo 2 to give the specified output bit. Thus, for example, 41, the first entry in the second set of parentheses, specifies the tap connections that produce the first bit of the second of the two parallel output symbols. The digit 4 means that there is a tap connection from bit position three of the first or left-hand stage of the SR, and the digit 1 implies that there is a tap connection from bit position one of the second stage of the SR.

The equivalent binary constraint length of a dual-*m* code is 2*m* data bits. This is consistent with the conceptual definition for binary codes, in which constraint length is the maximum number of channel bits that can be influenced by one data bit. Using this definition of constraint length, dual-*m* codes are inferior to binary codes of the equivalent constraint length by about 3–4 dB in E_B/N_0 at $P_B = 10^{-4}$.

Chapter 7
Threshold Decoding

7.1 INTRODUCTION

Threshold, or majority-logic, decoding first received significant attention and rigorous analysis in the doctoral thesis of J.L. Massey [7.1]. While Massey studied techniques for both convolutional and block codes, the treatment that follows considers threshold decoding of convolutional codes only. Threshold decoders were the first commercially produced decoders of convolutional codes. The past decade, however, has seen Viterbi decoding become the dominant technique for convolutional codes on the strength of its highly satisfactory bit error probability performance, speed of operation, ease of implementation, and low cost. Actually, threshold decoding possesses the last three of these attributes, too, but its bit error performance as a function of code constraint length is inferior to that of either Viterbi or sequential coding.

Of the three techniques just mentioned for decoding convolutional codes, threshold decoding is conceptually and practically the closest to block decoding. The conceptual similarity stems from the fact that threshold decoding of convolutional codes begins with the calculation of a set of syndrome digits, defined for convolutional codes in a manner very similar to their definition for block codes. This causes threshold decoding to lack the search aspects characteristic of Viterbi decoding and sequential decoding. The role of the syndrome sequence also makes implementation somewhat similar to that for decoding block codes, with the exception of the threshold element and the special type of parity checks involved.

The early popularity of threshold decoding as an error-control technique was due in part to this simplicity and in part to the fact that Massey's published results included several sizable families of suitable convolutional codes of various rates and structures. While these original codes and their decoding algorithm were designed to handle random errors, refinements such as diffuse threshold decoding (see Chapter 10 for details) were soon developed to cope with bursts as well.

As mentioned above, certain classes of reasonably good block codes have also been found to be threshold, or majority-logic, decodable. In fact, the earliest published work on majority-logic decoding was Reed's algorithm [7.2] for decoding Muller's block codes.

Threshold decoding of convolutional codes requires the calculation of a syndrome at the receiver in a manner similar to that used for block codes. Recall that in the block code case the received vector, which consists of information digits and check digits all possibly corrupted by noise, is operated on by the parity-check matrix to produce a syndrome vector **s**. With **x**, **y**, and **e** representing transmitted, received, and error vectors respectively, you saw that $\mathbf{s} = \mathbf{y}H = (\mathbf{x} + \mathbf{e})\,H = \mathbf{e}H$, since $\mathbf{x}H = 0$. Each multiplication of a vector by a column of H produces one scalar parity-check equation; i.e., one equation relating specific information digits and one more specific check digits by means of a modulo-2 sum. For the received vector, each multiplication by a column of **H** produces one digit of **s**. If $\mathbf{e} = \mathbf{0}$ or has an even number of errors in the positions checked by a column of **H**, then the corresponding component of **s** is 0. Note that what the syndrome calculation does is to recompute check digits on the basis of the received information digits and add each of these recomputed checks to the corresponding received check. If the corresponding computed and received check digits are the same, the result, a component of **s**, is 0.

In the case of convolutional codes the syndrome definition is similar, except that the syndrome is a sequence because the information and check digits occur as sequences. This is illustrated in the portions of Figure 7.1 labeled encoder and syndrome calculation.

For the explanation that follows, assume a rate $\frac{1}{2}$ code and modify slightly the notation used in conjunction with Figure 6.1. At the transmitter, let $i_0, i_1, i_2, \ldots$ represent information digits, and let $c_0, c_1, c_2, \ldots$ represent corresponding check digits. Let i_j' and c_j' be the received digits (possibly in error) corresponding to i_j and c_j, respectively. Finally, let e_{Ij} and e_{cj} be the error digits at the jth position in the information and check sequences, respectively, giving

$$\begin{aligned} i_j' &= i_j + e_{Ij} \\ & \qquad\qquad\qquad j = 0, 1, 2, \ldots \\ c_j' &= c_j + e_{cj} \end{aligned} \tag{7.1}$$

with all elements and operations from GF(2).

To understand how the system of Figure 7.1 works, start with $j = 0$ and assume that all shift register stages have initially been set to 0. Let a double prime denote a check digit calculated at the decoder.

Then

$$c_0'' = g_0 i_0' \qquad s_0 = c_0'' + c_0'$$

$$c_1'' = g_1 i_0' + g_0 i_1' \qquad s_1 = c_1'' + c_1'$$

$$c_2'' = g_2 i_0' + g_1 i_1' + g_0 i_2' \quad s_2 = c_2'' + c_2' \tag{7.2}$$

If the encoder SR is three stages long (i.e., $K = 3$), then subsequent calculations are

$$c_3'' = g_2 i_1' + g_1 i_2' + g_0 i_3' \qquad s_3 = c_3'' + c_3'$$

$$c_4'' = g_2 i_2' + g_1 i_3' + g_0 i_4' \qquad s_4 = c_4'' + c_4' \tag{7.3}$$

et cetera.

In threshold decoding, with the exception of one special case, the syndrome digits are not usually used directly to perform decoding. Rather, certain linear combinations of the syndrome digits are formed and these are used in conjunction with a majority logic element to provide an estimate $\hat{e}_{Ij}$ of the information error digit e_{Ij}. The threshold decoding process itself cannot be completely understood without first understanding the meaning, utility, and construction of these special linear combinations of the syndrome digits.

One final point should be noted before we discuss orthogonal parity checks in more detail in Section 7.2: Because each received information digit i' is corrected directly through use of the relationship:

$$\hat{i} = i' + \hat{e}_I,$$

it is clear that only systematic convolutional codes can be used with the decoding technique just described. (Here the subscript j has been suppressed and the circumflex (^) denotes an estimated value.)

7.2 ORTHOGONAL PARITY-CHECK EQUATIONS

The key concept in threshold decoding of convolutional codes is that of parity-check equations which are *orthogonal on a digit.* This concept of orthogonality differs from the well-known one from mathematics and signal theory in which two vectors or two functions are orthogonal to each other if and only if their inner or dot product, appropriately defined, is equal to zero. Rather, orthogonality of parity-check equations is defined as follows: Given a set of equations that relate various information and check digits through checksums, these equations are *orthogonal on digit* d_j if d_j is

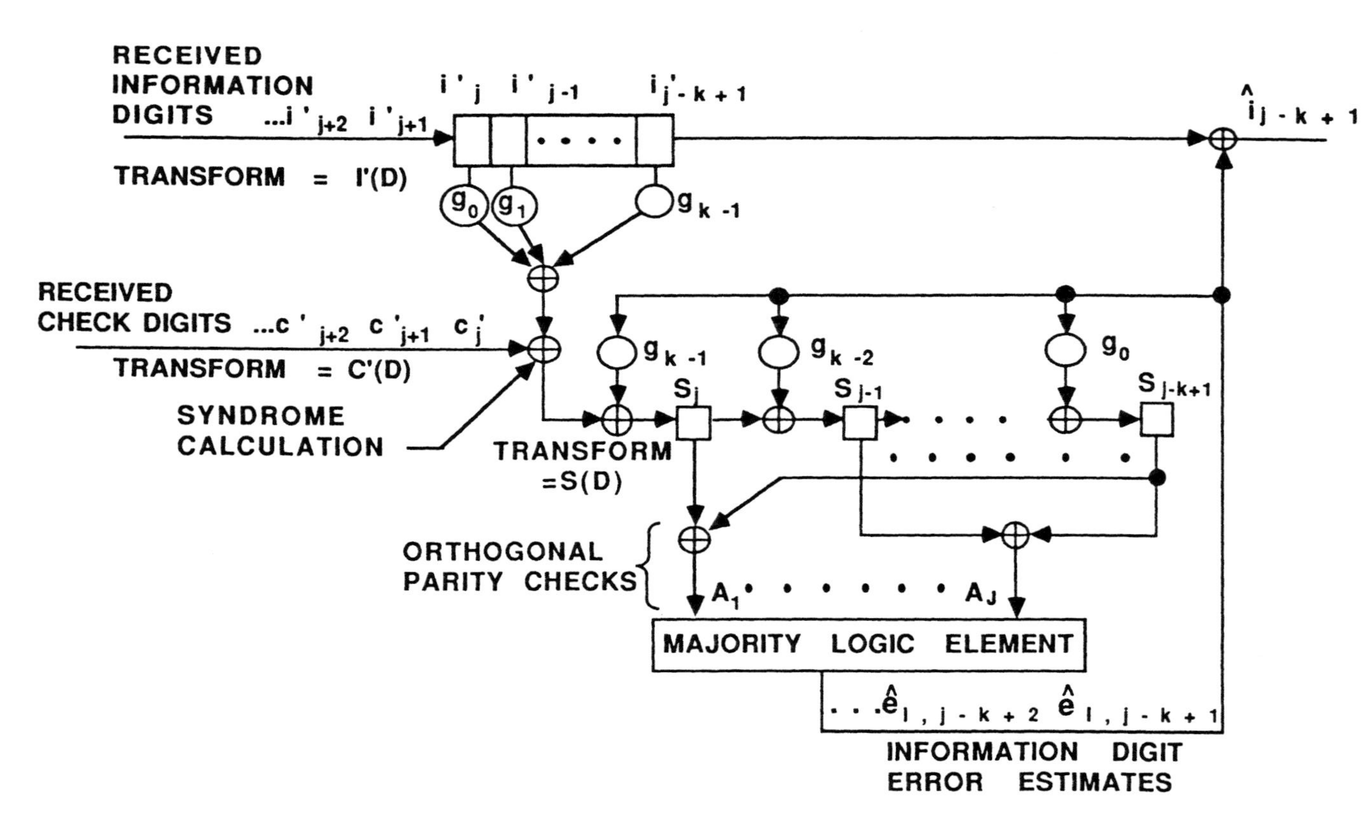

Figure 7.1 Majority-logic threshold decoder for rate $\frac{1}{2}$ binary systematic code.

checked by every equation and if no other digit is checked by more than one equation. Note that in order to achieve orthogonality it may be necessary to combine two or more of the original syndrome digit equations. (This process will be demonstrated in Example 7.2.) An additional goal while constructing orthogonal sets of equations is to maximize the total number of digits checked.

Example 7.1. Denote information digits by i, check digits by c, and syndrome digits by s. Then the set of equations:

$$\begin{aligned} s_1 &= i_1 + i_2 + c_1 + c_3 \\ s_2 &= i_1 + i_3 + c_2 + c_5 \\ s_3 &= i_1 + c_4 \end{aligned} \tag{7.4}$$

is orthogonal on i_1.

If equations (7.4) are the complete set of orthogonal equations for a code, then the code is called *self-orthogonal* because orthogonality has been achieved in terms of single syndrome digits without the need for combining them.

At this point, using the following notation results in both compactness and additional insight; let

$$\begin{aligned} I(D) &= i_0 + i_1 D + i_2 D^2 + \ldots \\ I'(D) &= i'_0 + i'_1 D + i'_2 D^2 + \ldots \\ E_I(D) &= e_{I0} + e_{I1} D + e_{I2} D^2 \ldots \end{aligned} \tag{7.5}$$

be the transforms of the transmitted information sequence, received information sequence, and received information error sequence, respectively, so that

$$I'(D) = I(D) + E_I(D) \tag{7.6}$$

Similarily, let $C(D)$, $C'(D)$, and $E_c(D)$ be the transforms of the transmitted and received check sequences and the received check error sequence, respectively, so that

$$C'(D) = C(D) + E_c(D) \tag{7.7}$$

Now, since information and check sequences are related by

$$C(D) = G(D)I(D), \tag{7.8}$$

the syndrome calculation at the decoder produces (referring to Figure 7.1 and using (7.6) and (7.7)):

$$\begin{aligned} S(D) &= G(D)I'(D) + C'(D) \\ &= G(D)[I(D) + E_I(D)] + C(D) + E_c(D) \\ &= G(D)E_I(D) + E_c(D) \end{aligned}$$

Note here that, since syndrome digits are actually linear combinations of digits of the error sequences, the notion of orthogonality applies to the error digits in exactly the same way that it does to the information and check digits.

Example 7.2 This is a systematic code from Massey's set of "trial-and-error" codes. It has $R = \frac{1}{2}$ and $k_{n_0} = 12$, with four orthogonal check equations. The generator polynomial for the check digits is $G(D) = 1 + D^3 + D^4 + D^5$. The complete decoder is shown in Figure 7.2. For the moment, look only at the encoder and syndrome portions. Figure 7.2 shows the calculation of s_5.

The syndrome digits are given (with all addition modulo 2) by

$$\begin{array}{lllllllllllll} s_0 = e_{I0} & & & & & & + e_{c0} & & & & & & \\ s_1 = & e_{I1} & & & & & & + e_{c1} & & & & & \\ s_2 = & & e_{I2} & & & & & & + e_{c2} & & & & \\ s_3 = e_{I0} & & & + e_{I3} & & & & & & + e_{c3} & & & \\ s_4 = e_{I0} & + e_{I1} & & & + e_{I4} & & & & & & + e_{c4} & & \\ s_5 = e_{I0} & + e_{I1} & + e_{I2} & & & + e_{I5} & & & & & & + e_{c5} & \end{array} \qquad (7.9)$$

These six equations are obviously not orthogonal on e_{I0}. To achieve orthogonality on e_{I0} it is necessary to examine various sums of equations as Massey did, with the additional goal of maximizing the number of orthogonal equations. Clearly s_1, s_2, and $s_1 + s_2$ are of no help since they do not involve e_{I0}, while s_0 and s_3 are acceptable as they are. This leaves s_4 and s_5 for consideration. To use them both, e_{I1} must be eliminated from one of them without involving e_{I2} more than once. It is also desirable to avoid eliminating e_{I2} entirely. There are several ways to accomplish this, as trial and error will show, but the solution chosen by Massey is the simplest: use s_4 and $s_1 + s_5$. The resulting set of equations is orthogonal on e_{I0} and checks all digits except e_{c2} of the maximum of 12 possible. Denote each member of a set of orthogonal parity checks by A_j. Then in this case the result is

$$A_1 = s_0 \quad A_3 = s_4$$
$$A_2 = s_3 \quad A_4 = s_1 + s_5 \tag{7.10}$$

7.3 THE THRESHOLD DECODING ALGORITHM

Remember that each received digit is actually given in (7.1) as

$$i'_j = i_j + e_{Ij}$$
$$c'_j = c_j + e_{cj}$$

and that each expression for an s_j holds with e_{Ij} replacing i'_j and e_{cj} replacing c'_j. The goal then is to estimate correctly the single noise digit e_{I0}. Call this estimate $\hat{e}_{I0}$, and let $\hat{i}_0$ be the estimate of i_0. Then $\hat{i}_0 = i'_0 + \hat{e}_{I0} = (i_0 + e_{I0}) + \hat{e}_{I0}$. Thus, as expected, i_0 is correctly decoded (i.e., $\hat{i}_0 = i_0$) if e_{I0} is correctly estimated. Following Massey, let J represent the total number of parity-check equations orthogonal on i_0. Then the majority decoding rule is: Set $\hat{e}_{I0} = 0$ if one-half or more of the orthogonal check equations are equal to zero; otherwise set $\hat{e}_{I0} = 1$. This rule results in correct decoding as long as the error-correcting capability of the code has not been exceeded. According to [7.1], for majority decoding, $\lfloor J/2 \rfloor$ or fewer errors among the digits actually checked by the orthogonal equations always permit correct decoding of i_0.

How does this capability relate to d_{min} for a code? It turns out that if $d_{min} - 1$ check equations orthogonal on e_{I0} (or i_0) can be formed, then i_0 will be correctly decoded for any error pattern of weight $\lfloor (d_{min} - 1)/2 \rfloor$ or less. Nearly all of the trial-and-error codes found by Massey [7.1] meet this condition.

The general majority-logic threshold decoder for the BSC has the block diagram shown in Figure 7.1. Note the portion marked decision feedback. This concept, not mentioned thus far, involves picking off the estimates $\hat{e}_{Ij}$ of the information digit errors and feeding them back through a parity network based on the parity network used in the encoder, and then adding the output digits of this feedback parity network into corresponding stages of the syndrome register. This feedback is carried out for the purpose of removing from the digits in the syndrome register the effects of the error in the information digit just decoded. If the error estimate is correct, then the effect of the feedback is to provide a syndrome that "acts" as if there were no prior errors in the information sequence, identical to the initial conditions when i_0 was being decoded. If, however, the error estimate is wrong, then obviously not only is a decoding error committed

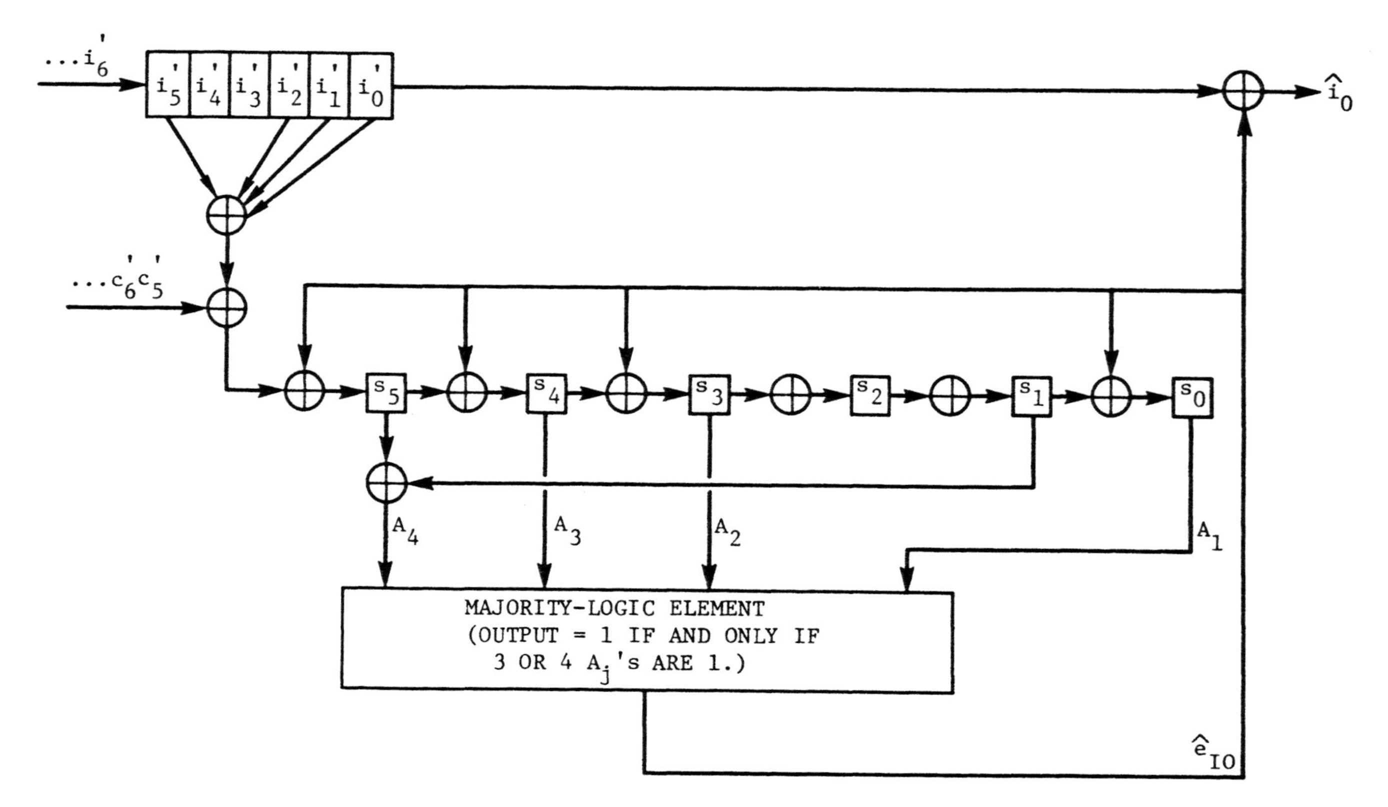

Figure 7.2 Decoder for Example 7.2.

but incorrect information is fed back to the syndrome. This incorrect feedback usually has the effect of causing decoding errors where none may have been committed in its absence. This so-called *error-propagation* effect is probably the most serious flaw in algorithms using convolutional codes. The effects and prevention of error propagation are discussed further in Section 7.4.

To understand how decision feedback works (or doesn't work, in the case of an incorrect decision), consider once again the rate $\frac{1}{2}$ code of Example 7.2. The essential code parameters are repeated here for convenience:

$$G(D) = 1 + D^3 + D^4 + D^5$$
$$A_1 = s_0,\ A_2 = s_3,\ A_3 = s_4,\ A_4 = s_1 + s_5$$

Thus,

$$\begin{aligned}
A_1 &= e_{I0} && && && + e_{c0} \\
A_2 &= e_{I0} && + e_{I3} && && + e_{c3} \\
A_3 &= e_{I0} + e_{I1} && + e_{I4} && && + e_{c4} \\
A_4 &= e_{I0} + e_{I2} && + e_{I5} + e_{c1} && && + e_{c5}
\end{aligned} \tag{7.11}$$

Consider first the expression:

$$s_0 = e_{I0} + e_{c0}$$

If $e_{I0} = 0$, there is nothing for which to compensate. If, however, $e_{I0} = 1$, then its effect should be removed from the syndrome digits before computing the next set of values A_js. This can be done more easily if the (correct) estimate $\hat{e}_{I0} = 1$ is added to s_0 to give $\tilde{s}_0 = s_0 + \hat{e}_{I0}$. Here the tilde ($\tilde{}$) indicates syndrome digits modified by decision feedback. More generally, whether $e_{I0} = 0$ or 1, addition to s_0 of the *correct* estimate $\hat{e}_{I0} = e_{I0}$, yields

$$\tilde{s}_0 = e_{I0} + \hat{e}_{I0} + e_{c0} = e_{c0} \tag{7.12a}$$

Similarly, this same correct feedback also yields

$$
\begin{aligned}
\tilde{s}_3 &= \quad\quad\quad e_{I3} \quad\quad\quad\quad + e_{c3} \\
\tilde{s}_4 &= e_{I1} \quad\quad\quad + e_{I4} \quad\quad\quad\quad + e_{c4} \\
\tilde{s}_5 &= e_{I1} + e_{I2} \quad\quad + e_{I5} \quad\quad\quad\quad + e_{c5}
\end{aligned}
\tag{7.12b}
$$

showing that the effects of e_{I0} have been completely removed. The next step in the decoding is to estimate e_{I1}. Before this is done, the encoder and syndrome SR contents are both shifted one stage to the right, "bumping" i_0' and $\tilde{s}_0$. The new contents of the syndrome SR (with Stage 1 at the right) are:

$$
\begin{aligned}
&\text{Stage 1:} \quad e_{I1} && + e_{c1} \\
&\text{Stage 2:} \quad e_{I2} && + e_{c2} \\
&\text{Stage 3:} \quad e_{I3} && + e_{c3} \\
&\text{Stage 4:} \quad e_{I1} + e_{I4} && + e_{c4} \\
&\text{Stage 5:} \quad e_{I1} + e_{I2} \quad e_{I5} && + e_{c5} \\
&\text{Stage 6:} \quad e_{I1} + e_{I2} + e_{I3} + e_{I6} && + e_{c6}
\end{aligned}
\tag{7.13}
$$

Equations (7.13) are identical to (7.9) except that all subscripts have been increased by 1. Note that Stage 6 concurrently has received the newest syndrome digit. Thus, the decoder can estimate e_{I1} in exactly the same way e_{I_0} was estimated.

7.4 ERROR PROPAGATION: ANALYSIS AND PREVENTION

What happens if $\hat{e}_{I0} \neq e_{I0}$? Look at the expressions for s_0, s_3, s_4, and s_5. In each case, the effect is to add 1 (modulo 2) to the correct expression, causing (in this case) these four digits to be in error as long as they remain in the syndrome register, unless changed by another error. In general, this incorrect feedback causes subsequent decoding decisions to be based on incorrect information in the syndrome register. This can be avoided, however.

The most obvious way is to break the feedback connection in Figure 7.2. This has the disadvantage of failing to feed back correct information, which shows up as increased probability of decoding error. There is, however, never any error propagation. This type of decoding has been termed *direct decoding* by Robinson [7.3].

A second way is to design the code to prevent all but a minimal (2 to 3 constraint lengths) propagation of decoding errors. The worst situation of this sort is catastrophic error propagation, in which the decoder continues to make errors even when its syndrome inputs are all 0s, signaling no transmission errors. This problem and its solution are discussed later in this section.

A third way, suggested by Massey [7.1], is to count errors. When the decoder appears to be correcting errors beyond its design capability by producing a great many 1s as error estimates, it must actually be propagating one or more decoding errors. The receiver can then signal the transmitter to restart at some time in the past, or it can clear itself out and start anew without any retransmission.

As you just saw, the use of decision feedback to remove the effects of received errors in information digits can lead to a serious problem if the decoding decision is incorrect, because the wrong information is being fed back. At the very least, this incorrect information results in error propagation; that is, making additional errors that would not have been made with correct feedback or with no feedback at all. In spite of this error propagation, subsequent receipt of error-free channel digits usually causes the syndrome register to return to the all-zero state because all syndrome inputs are zero. Sometimes, however, incorrect decision feedback causes a majority of ones to occur repeatedly in the syndrome even with zeros as input. In this case the decoder continues to decide that there are errors in the received information digits. This is catastrophic propagation of decoding errors, in which bits can be decoded in error indefinitely.

Massey and Sain [7.4] showed that certain properties of code generators inevitably result in catastrophic error propagation, while other properties assure the lack of error propagation. The test is a simple one: Catastrophic error propagation is possible if and only if the code generators, written in the usual delay-operator notation, contain a common factor of degree one or greater. An important class of codes that do not suffer from catastrophic error propagation is the set of all systematic codes, since one of the generators is simply $G(D) = 1$.

Example 7.3. The code with generators $1 + D$ and $1 + D^2$ permits catastrophic error propagation, because $1 + D^2 = (1 + D)^2$ (modulo 2). On the other hand, the code with generators $1 + D$ and $1 + D + D^2$ does not allow catastrophic propagation.

7.5 PERFORMANCE ASSESSMENT

Threshold decoders are simple to implement, and they operate at a constant output rate. Aside from error propagation, which you just learned is a solvable problem, the chief disadvantage of the threshold decoding technique for convolutional codes is that bit error probability is asymptotic to value greater than zero as constraint length increases, a flaw not shared by Viterbi and sequential decoding. Nonetheless, the performance of threshold decoding is still adequate for many situations. Although Viterbi and sequential decoding of convolutional codes offer superior P_B performance compared to threshold decoding, the latter should still be considered for some situations such as channels characterized by bursts of errors. (See Chapter 10.)

Figure 7.3 shows the probability of first decoding error P_B as a function of K for $R = 1/2$ and $R = 1/3$ trial-and-error codes over a binary symmetric channel with a range of channel bit error probabilities. These curves were plotted from data contained in Figures 16 and 17 and Table II of [7.1]. The lack of exponential decay of P_B as a function of K is quite evident. (The curves would be straight lines for the axis scales used if the decay were truly exponential.) On the other hand, for the range of values of K shown, the asymptotic lower bound to P_B is evident only for channel bit error probabilities p of 0.0615 and 0.0310 at $R = 1/2$, and for $p =$ 0.1100, 0.0795, and 0.0615 at $R = 1/3$. This suggests that for constraint lengths K of about 20 or less, the failure of P_B to go to zero may be a problem only for relatively poor channels (i.e., $p > 0.01$).

The data of Figure 7.3 have been replotted in Figure 7.4 in a more familiar form showing P_B as a function of p, with K as a parameter.

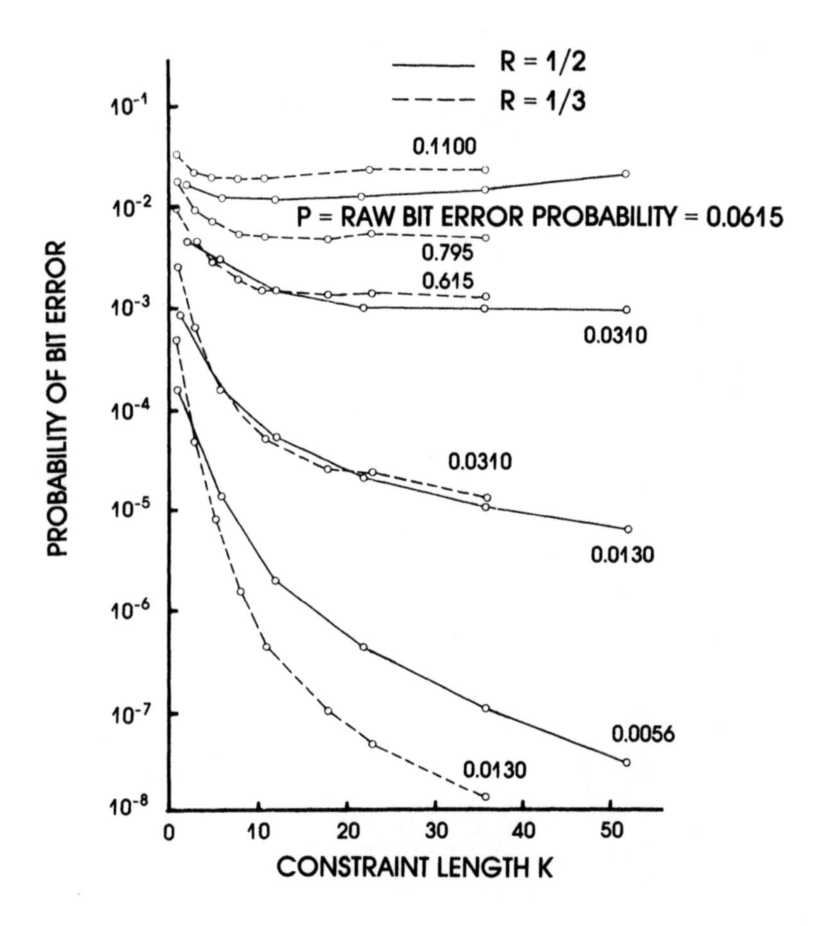

Figure 7.3 Performance of trial-and-error codes with majority decoding on the BSC as a function of constraint length.

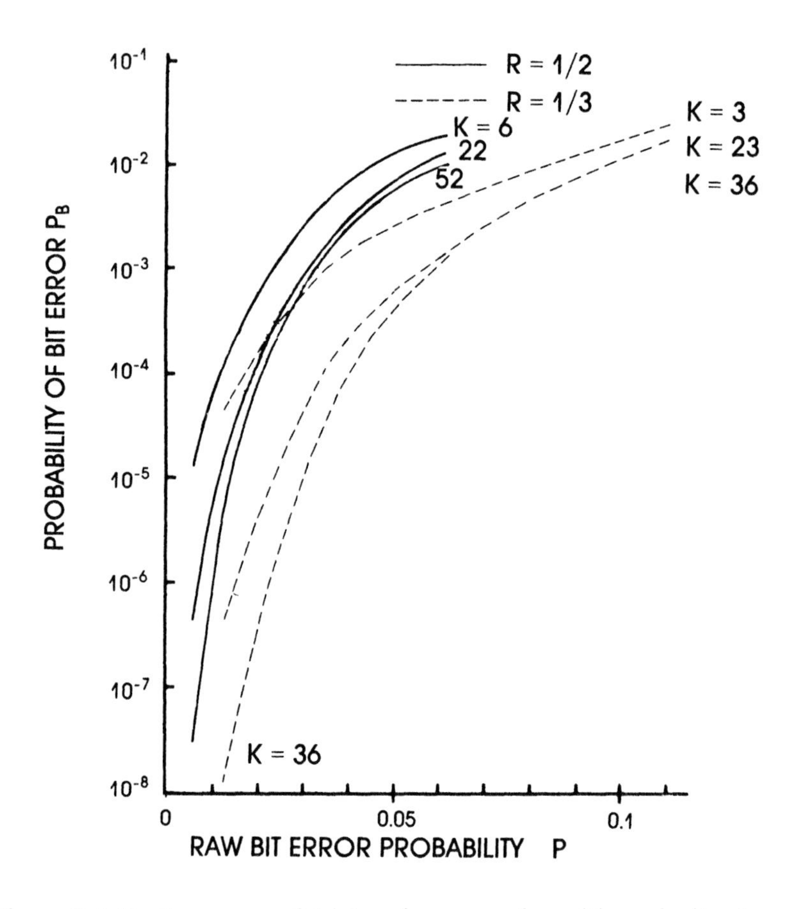

Figure 7.4 Performance of trial-and-error codes with majority decoding on the BSC as a function of channel error probability.

Chapter 8

The Viterbi Decoding Algorithm

8.1 INTRODUCTION

The Viterbi [8.1] decoding algorithm is a powerful, fast, and easily implemented technique for decoding convolutional codes. Furthermore, its analysis and simulation are easily carried out. The amount of decoding computation needed is independent of the noise in the received sequence of channel symbols.

The R = 1/3, K = 3 systematic convolutional code and encoder introduced in Chapter 6 will again be used as an example. That encoder is reproduced for convenience in Figure 8.1. It will also be very useful to have at hand the corresponding trellis. This is shown again in Figure 8.2. In both figures, the oldest symbols are at the left, as they were in Chapter 6. The reason for showing the trellis once more is that the Viterbi algorithm is based on finding the most likely path, in a sense to be defined shortly, through the trellis. Remember that the trellis exploits the fact that there is no dependence between encoded symbols that lie more than one constraint length from each other.

For instance, if 1101 . . . is a bit sequence entering the encoder of Figure 8.1, then the first three bits of the input sequence produce the following encoder output: 111 100 010. (See Figure 8.2.) Now shift in the fourth bit, placing 101 in the encoder. The first encoder output (110) with this latest bit in the encoder is independent of whether the very first input bit was a 0 or a 1. Stated another way, the last node reached in the trellis (arrow, Figure 8.2) could have been reached by any of the following input sequences: 1101, 1001, 0101, and 0001. Recall that in the SR state representations of Figure 8.2, the contents of Stage 1 are shown to the left of the Stage 2 contents.

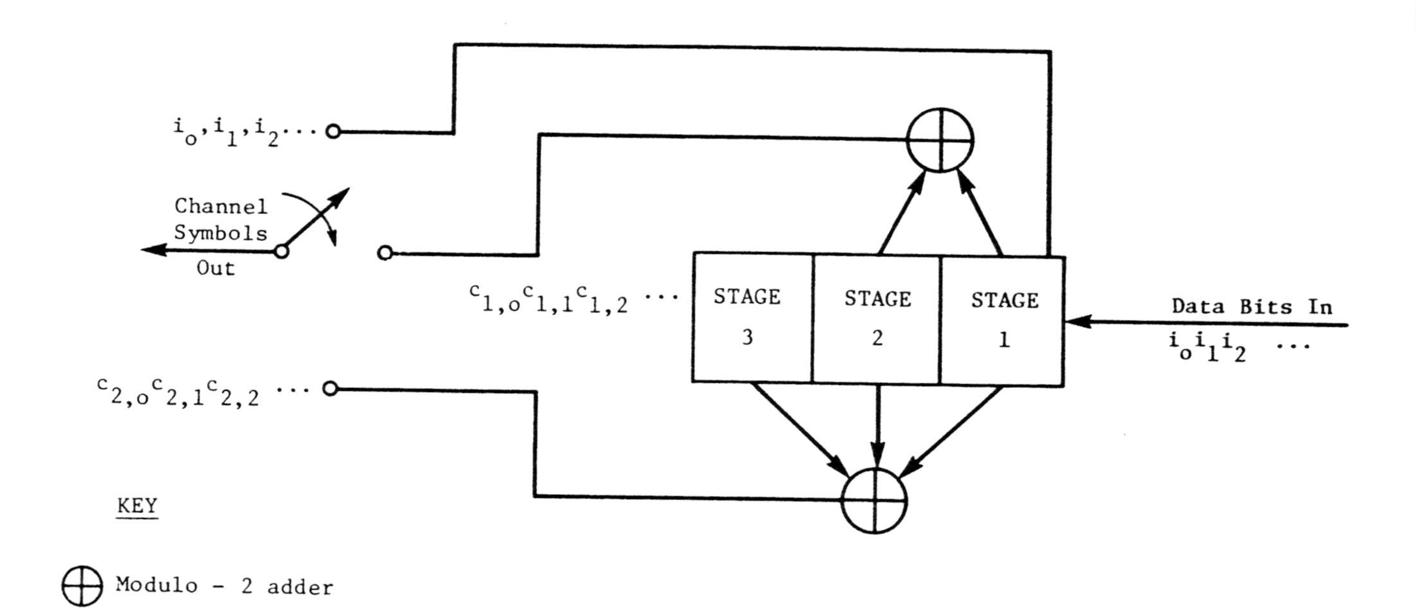

Figure 8.1 Convolutional encoder.

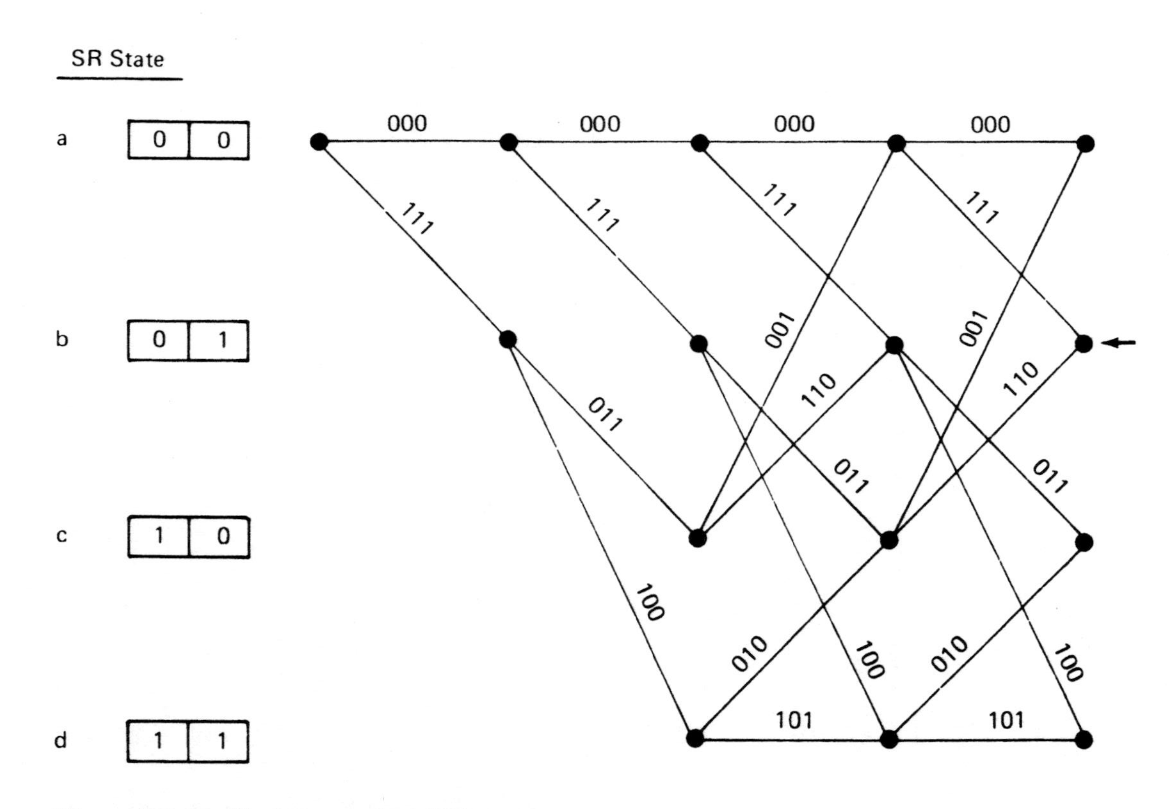

Figure 8.2 Trellis for encoder of Figure 8.1.

As with block codes, the goal is to find, from among all possible transmitted sequences, the one that is the best match in some sense to the received sequence. The initial discussion is based on the binary symmetric channel. Later in this chapter (Section 8.5), up to 8-level (3-bit) quantization of the channel (i.e., demodulator) output will be discussed because of the resulting decoder performance improvement of about 2 dB relative to the performance using hard decisions (1-bit quantization).

8.2 DECODER OPERATION

Viterbi decoding, in common with other decoding techniques, involves choosing that trial information sequence the encoded version of which is in some sense closest to the received sequence. For the time being, Hamming distance will be used to measure of closeness or similarity of two binary sequences.

Suppose that the all-0s sequence were transmitted and the following sequence were received (oldest symbol at the left):

010 000 100 001

Now try to decode, that is, correctly estimate the transmitted sequence using the trellis of Figure 8.2 and Hamming distance, keeping in mind that a systematic code is being used. The trellis has been redrawn in Figure 8.3 to display the distances between corresponding branches of the trial and received sequences, with the latter shown below the trellis. A circled number at a node is the cumulative distance between the received sequence and one of the paths from the base of the trellis out to that node. When there are two numbers, the first number corresponds to the upper path into that node. For example, the first branch received, 010, is distance 1 from 000 and distance 2 from 111.

The distances between the second branch of the received sequence and each of the corresponding four possible trial sequence branches are easily seen to be 0, 3, 2, and 1 for the branches 000, 111, 011, and 100, respectively. Thus, the four possible transmitted sequences of length six channel digits are at distance 1, 4, 4, and 3 from the first six received digits.

Up to this point ($2 = K - L$ branches), there has been only one possible way to reach each node on the path, and all possible transmitted sequences were tested. The next (third or Kth) level node is the first time a node can be reached by more than one path. This leads to the possibility of two distinct distances to that node, one for each of the two unique paths

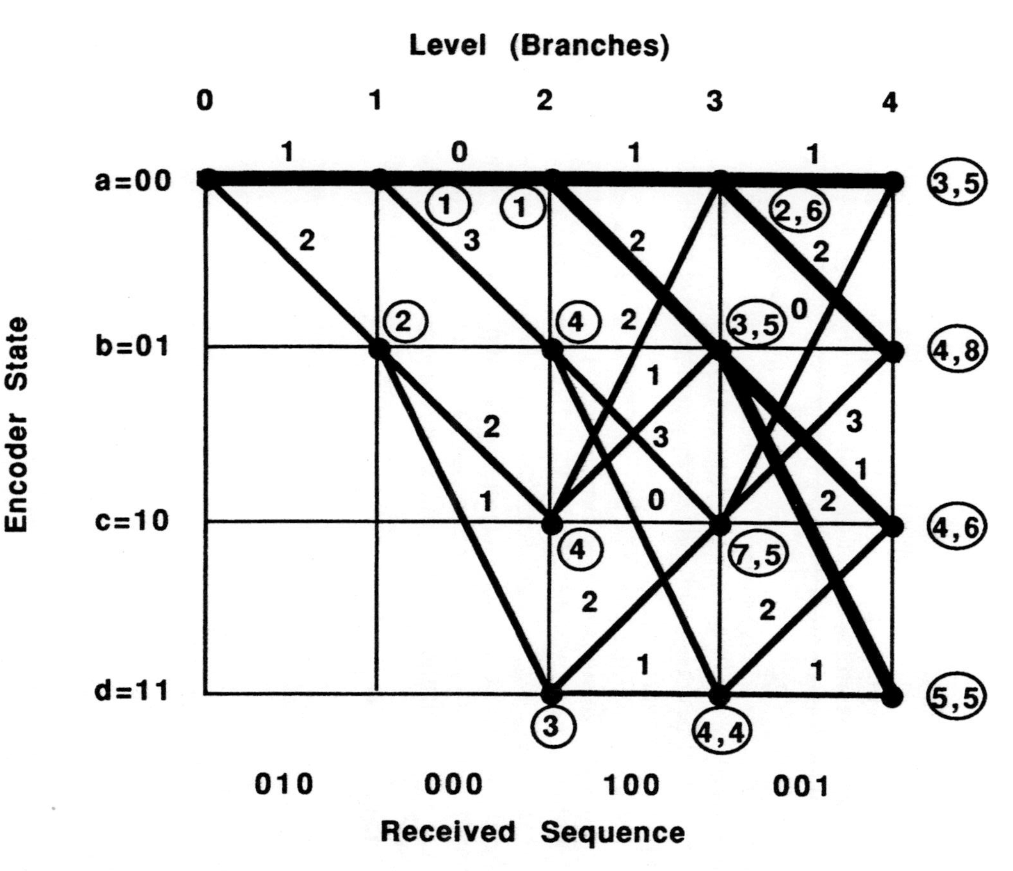

Figure 8.3 Code trellis showing branch and path Hamming distances.

leading to it from the base of the trellis. Here, for the first time, the decoder must decide which path to retain for subsequent extensions. If this choice were not made at the Kth and each later level, the number of paths would grow exponentially with message length.

The above discussion motivates the key step in Viterbi decoding. Beginning with the Kth branch and continuing for as long as there are data on which to operate, the decoder retains for each node only one of the two branches leaving that node. The branch retained is the one that gives the better overall path distance with respect to the received sequence. This process of discarding one of the two branches at each level from the Kth onward holds constant at 2^{K-1}, the number of paths that are extended to the next level. Viterbi [8.1] showed that this process is optimum in the sense of producing a maximum likelihood decision as to the correct path. For our example, a look at Figure 8.3 shows that the paths indicated with heavy lines will be retained in going from level 3 to level 4. Viterbi called these paths *survivors*.

At the end of three branches, Figure 8.3 shows that encoder state d corresponds to two survivors, the two distance-4 paths, 000 111 100 and 111 100 101. Which one should you keep? It turns out that it doesn't matter! In either case, the path distances from this node to the next level will be the same. Therefore, you could either flip a coin or you could arbitrarily choose a particular branch (say the upper) into such a node.

Consider what the decoder has accomplished so far. At level 3 there are four survivors (candidate-transmitted sequences) with distances as shown below:

Sequence	*Distance from received sequence*
000 000 000	2
000 000 111	3
111 100 010	5
000 111 100	4

It is interesting to note that the sequence having the best distance turns out (in this case) to be the one actually transmitted. Furthermore, the three best distances belong to the three paths that begin with the correct transmitted initial branch.

Now continue the decoding process. The terminal branch of each survivor at level 4 is shown in Figure 8.3 as a heavy line. Note that survivors at states c and d at level 3 do not survive to level 4. Rather, each of the level-3 survivors at states *a* and *b* produces two survivors to level 4. At this stage, the details are

Survivor	*Distance*
000 000 000 000	3
000 000 000 111	4
000 000 111 011	4
000 000 111 100	5

Note that now all four survivors happen to be correct not only in the first but also the second transmitted branch. In fact, the level-4 survivors at states *a* and *b* are the correct path over the first constraint length.

The brief tables in the two preceding paragraphs show that there are two kinds of quantities that must be "remembered" by the decoder: the surviving path and its distance from the received sequence for each node at each level in the trellis.

The process just described can now be repeated over the remainder of the trellis.

The next question is: how can you terminate the decoding process and make decisions (i.e., best estimates) as to the transmitted data bits? One way is to produce a decision at each level of the trellis, starting with the *K*th level. In this case, the obvious output is the oldest bit in the trellis path having the smallest distance from the received sequence. Once the decision has been made, the decoder can delete from its memory the 2^{K-1} branches at the position of the decoded bit in the trellis. The usual way of making decoding decisions, however, is to withhold any decision until the entire received sequence has been processed and then output the decoder's estimate of the entire message.

At the encoder, the final data bit is followed by a "flush" of $K - 1$ 0s. The decoder also knows the message length, so it knows when to truncate the trellis corresponding to this flush.

The final maximum likelihood path through the trellis can be quickly and uniquely identified by starting from the right and working backwards through the trellis. You can see that this procedure immediately chooses the best path into each mode and eliminates the need for any "look ahead" to determine the optimum overall path estimate. In terms of Figure 8.4, there would almost always be only one best path into state *a* at level 6. Proceeding leftward along this path would lead uniquely to a state at level 5, from which a best choice back to level 4 would be made, *et cetera*.

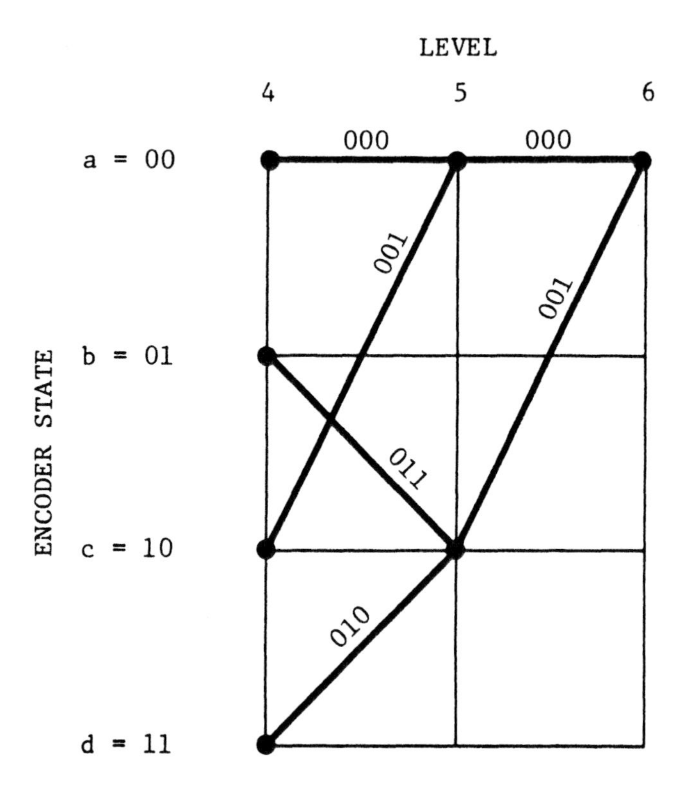

Figure 8.4 Truncation of trellis forced by all-zeros input.

Finally, note that while this example involved a systematic code, a nonsystematic code is just as easy to use because both encoder and decoder already know whether each branch in the trellis was generated by a 1 or a 0 into the encoder SR.

8.3 DECODER PARAMETERS

The preceding discussion should have made it clear that the decoder progresses at a fixed rate through the trellis. For real-time operations, a speed advantage of about 2^K relative to the channel rate is necessary because the decoder must process 2^K branches at each stage, discarding half of them. Depending on the decoding delay, there must be storage for some multiple of $M \times 2^{K-1}$ pairs of branch and path distance values, where M is the number of data bits in the message. These requirements resulted in a maximum K of 7 for many implementations. Recently, however, developmental Viterbi decoders with $K = 9$ or 10 have been reported, reflecting further advances in integrated circuit technology.

8.4 SOFT DECISIONS

Soft decisions on demodulator outputs result when additonal information is retained beyond that needed to make a 0/1 bit decision. For example, the 0 axis in Figure 2.2 could be divided into four or eight regions rather than the two regions shown. (The number of regions is usually a power of 2.) The presence of the demodulator output in any region then serves as an indication of the quality of that output. This quantization of the demodulator output space is implemented by using two or more bits to specify a value of 0 located somewhere in the middle of the region, or far from the origin in the case of large positive or negative amplitudes. An ideal (noise-free) demodulator output value must also be specified for each input. Clearly, Hamming distance is no longer applicable. Instead, it can be shown that minimizing a sum of squares of differences between ideal and received (quantized) outputs results in maximum likelihood decoding.

The use of soft decisions affects both storage and speed requirements at the decoder. The storage is affected only slightly in terms of two or three bits rather than one bit per metric value for each branch or path. Decoder speed must be greater to handle the subtraction and sum-of-squares computations in place of the exclusive-OR and bit counting needed for Hamming distance.

8.5 SIMULATION AND IMPLEMENTATION RESULTS

Viterbi decoding has been extensively simulated both as research effort and by organizations that manufacture and sell Viterbi decoder hardware for sale.

The results that follow were obtained by one organization engaged in the building of error-control hardware. A rate 1/2, nonsystematic, binary convolutional code used on a Gaussian channel was simulated using both hard (1-bit quantization) and soft (3-, 4-, and 5-bit quantization) decisions on channel digits. Constraint lengths K of 5, 6, and 7 were used, along with decoding delays of 3, 4, 5, and in a few instances, as many as 10, constraint lengths. Results were presented in terms of data bit error probability (P_B) as a function of the ratio of data bit energy to noise power density (E_B/N_0). Comparisons are made at $P_B = 10^{-5}$, which is frequently used as a baseline P_B. The results are presented in Tables 8.1, 8.2, and 8.3.

Comparison of Tables 8.1 and 8.2 shows that

1. Some improvement in E_B/N_0 is achieved by increasing constraint length with both hard and soft decisions.
2. Improvement with increasing decoding delay is moderate in going from 3 to 4 constraint lengths, with only marginal improvement beyond 4.
3. Significant improvement takes place with 3-bit quantization compared with hard decisions at any constraint length and decoding delay.

Finally, Table 8.3 shows that there is only marginal improvement in E_B/N_0 with quantization beyond 3 bits. It is useful, also, to observe the coding gain, which is the improvement in E_B/N_0 obtained with coding compared with ideal (coherent) PSK and no coding at $P_B = 10^{-5}$. You can easily infer that the latter requires $E_B/N_0 = 9.6$ dB to give $P_B = 10^{-5}$. Knowing this, you could compute the coding gain associated with each of the data points in Tables 8.1 and 8.2.

Table 8.1 E_B/N_0 (dB) for Hard-Decision Viterbi Decoding ($P_B = 10^{-5}$) [8.2]

Decoding Delay, L (*Constraint Lengths*)	*Code Constraint Length*		
	5	6	7
3	7.4	7.2	6.7
4	7.25	6.8	6.5
5	7.2*	6.7*	6.4
10	–	–	6.3

*$L \geq 5$

Table 8.2 E_B/N_0 (dB) for Soft-Decision (3 bits) Viterbi Decoding ($P_B = 10^{-5}$) [8.2]

Decoding Delay, L (*Constraint Lengths*)	*Code Constraint Length*		
	5	6	7
3	5.7	5.2	5.1
4	5.3	5.0	4.7
5	5.2*	4.9*	4.5
27	–	–	4.4

*$L \geq 5$

Table 8.3 Coding Gain *versus* Bits of Quantization (Constraint Length = 6, $P_B = 10^{-5}$) [8.2]

Number of Bits of Quantization	E_B/N_0(dB)	*Coding Gain* (dB)
3	4.9	4.7
4	4.8	4.8
5	4.7	4.9

Data in Tables 8.1 through 8.3 are reprinted with permission from General Atronics, a Magnavox subsidiary [8.2].

8.6 ASSESSMENT AND CONCLUSIONS

Section 8.3 showed that the decoder must keep track of 2^{K-1} paths through the trellis in order to identify each path on a branch-by-branch basis and to keep a complete record of all path metrics. Thus, storage and computation speed requirements both grow exponentially with constraint length. On the other hand, the algorithm's exponential decrease in bit error probability as a function of constraint length and the constant required decoder speed (independent of channel noise) often make it the preferred choice among convolutional code techniques for use on memoryless channels. Viterbi decoding alone does not perform well against bursts of errors, producing output error bursts, due to error propagation, of several times the input burst length. As discussed in Chapter 10, this deficiency can be removed with the use of interleaving. Equipment combining interleaving with Viterbi decoding is available commercially.

Chapter 9
Sequential Decoding

9.1 INTRODUCTION

Sequential decoding is a very powerful decoding technique for convolutional codes. Like Viterbi decoding it exhibits an exponential decrease in probability of bit error P_B as a function of code constraint length and is easy to implement for high-speed (Mb/s) operation with a manageable amount of hardware and software. Both memory and computation effort grow at rates that are less than exponential.

Sequential decoding for the BSC was invented and first studied by J.M. Wozencraft [9.1]. Major refinements and extensions were made by Fano [9.2], Reiffen [9.3], Jelinek [9.4], and Zigongirov [9.5]. Reiffen extended Wozencraft's theoretical results to the discrete memoryless channel (DMC) and improved upon the efficiency of the decoding algorithm. Fano generalized Wozencraft's and Reiffen's encoding scheme and refined and formalized the tree-searching technique used in decoding. Finally, Jelinek and Zigongirov, working independently of each other, developed what they have called a "stack algorithm," in which increased memory requirements are accepted in return for increased speed (relative to the Fano algorithm).

Conceptually, sequential decoding is quite simple, although the details may be somewhat involved. All of the techniques entail searching for a path through a message tree. This path, even though seldom correct over its entire length, has very high probability of being correct over the portion that includes the oldest digits, i.e., the information digits about to be decoded. The Fano and Jelinek-Zigangirov algorithms will be discussed here.

Sequential decoding differs fundamentally from decoding schemes developed for block codes and from other algorithms developed for convolutional codes. The basic task in sequential decoding involves searching the code tree to determine a trial information sequence the encoded version of which most closely matches the corrupted received sequence. As with all decoding, the notion of closely matching sequences involves the concept of distance. The simplest concept of distance (and the one used by Wozencraft [9.1]) is Hamming distance, which has already been used extensively in this book for block and convolutional codes. Other functions of the trial and received sequences that could be used as decoding metrics include the following: mutual information of a particular trial sequence and the received sequence (see, for example, Reiffen [9.2]); the sum of the squares of the differences between matched filter outputs for pairs of corresponding trial and received digits for a channel with a binary input space and a finely quantized output space (as in the simulation reported by Reiffen and Wiggert [9.6]).

9.2 A FUNDAMENTAL LIMIT ON SEQUENTIAL DECODING

Before discussing sequential decoding techniques, it is important to be aware of a fundamental limitation on sequential decoding in general. This limitation has a serious impact on the decoder speed required (especially in real time) to handle particularly noisy received sequences. You will see that such a received sequence usually results in a search of the message tree before the sequence is accepted by the decoder. While this search is taking place, received digits continue to enter the decoder. It has been possible to characterize this limitation quantitatively in terms of a parameter R_{comp}, the computational cutoff rate. It turns out that as the channel rate R approaches R_{comp} from below, the computational load increases tremendously. In fact, in theory at least, the average computational effort per decoded bit becomes infinite when $R = R_{comp}$. While this does not happen in practice (because the theoretical result is only a bound), the dramatic increase in computation, even as R exceeds about 0.9 R_{comp}, has certainly been observed.

The parameter R_{comp} is determined entirely by the channel output symbol energy to noise power density ratio. For the DMC with I input symbols, J output symbols, and symbol input probabilities p_i, R_{comp} is given by

$$R_{comp} = \max_{\{p_i\}} \left\{ -\log \sum_{j=1}^{J} \left[\sum_{i=1}^{I} p_i \sqrt{p(j|i)} \right]^2 \right\} \quad (9.1)$$

where the $p(j|i)$ are channel transition probabilities and the maximization takes place over all possible sets of input probabilities.

For the BSC, (9.1) reduces to

$$R_{\text{comp}} = 1 - \log\,[1 + 2\sqrt{p(1 - p)}], \tag{9.2}$$

while for the binary-input, continuous-output channel of Section 2.2, the result is

$$R_{\text{comp}} = \log\,[2/(1 + \exp(-E_S/N_0))], \tag{9.3}$$

where E_S is the received energy per channel digit, given by $E_S = E_B R$ for energy per data bit E_B and rate R and binary modulation. If logarithms to the base 2 are used, (9.3) becomes

$$R_{\text{comp}} = 1 - \log_2\,(1 + \exp(-E_S/N_0)) \text{ bits/symbol} \tag{9.4}$$

Thus, we see that R_{comp} depends on the average properties of the channel (through N_0) and on the received energy. Note that the received energy E_S actually depends also on properties of the transmission medium in ways that are beyond the scope of this book ($1/r^2$ attenuation, absorption, and dispersion, for example).

Next, we shall discuss sequential decoding algorithms in detail. Then we shall quantitatively examine how decoding computation depends on the ratio R/R_{comp}.

9.3 SEQUENTIAL DECODING: GENERAL TECHNIQUE

A replica of the encoder generates a candidate-transmitted (or trial) sequence k_0 information digits or n_0 channel digits at a time, starting at the base of the message tree. As each new set of n_0 digits is added to the existing sequence, the metric function for trial and received sequences is computed. This number is compared with a threshold that is a function of the length of the trial sequence and of the number of times the decoder has returned to the base of the tree during its search. This threshold is just one value in a family of thresholds for the present level of strictness in rejecting trial sequences. These thresholds can be visualized as shown in Table 9.1.

Table 9.1 Relationships among thresholds for sequential decoding.

Sequence Length (branches)	*Family* 1	2
.	.	.
.	.	.
.	.	.
3	$T_3^{(1)}$	$T_3^{(2)}$
2	$T_2^{(1)}$	$T_2^{(2)}$
1	$T_1^{(1)}$	$T_1^{(2)}$

Note: $T_i^{(j)} < T_{i+1}^{(j)}, < T_i^{(j)} < T_i^{(j+1)}$ for Hamming distance.

The decoder operates in one of two modes. A search mode is entered under either of two conditions: (1) at the initiation of decoding, and (2) any time the decoder fails to extend the trial sequence. When the decoder is successful in extending the trial sequence, operation continues in a steady state mode. Decoding a digit means that the decoder accepts a trial sequence that begins with that digit and is at least one constraint length long.

The decoder begins operation by generating all q^{k_0} branches emanating from the base of the tree and choosing the best of these in terms of its metric relative to the first received branch. In many applications, $q = 2$ and $k_0 = 1$, giving two branches leaving each node. From now on, we will assume these values. At the next level node, both branches are generated and tested using the next threshold from the same family. The branch giving the better value of the distance function is chosen. As long as the trial sequence satisfies the threshold, the decoder continues to add branches until the sequences are being compared over a full constraint length. At this point, each time the threshold is satisfied, k_0, information digits can be decoded, or the decoder can continue to operate on the entire sequence it has produced, extending it to some prespecified length before outputting any decoded digits.

Any time the trial sequence fails to satisfy the threshold, the decoder re-enters the search mode. If these failures continue, ever shorter sequences are tested, until at some point the decoder is back at the base of the tree. If this happens, the decoder must loosen the threshold and begin working its way out as if it were just starting up.

9.4 THE FANO SEQUENTIAL DECODING ALGORITHM

Fano's algorithm (see [6.3] or Chapter 6 of Wozencraft and Jacobs [2.1]) will work with any one of a variety of decoding metric functions, as described in Section 9.1. Perhaps the most common of these is the logarithm of the likelihood ratio:

$$\log \frac{P(\mathbf{z}|\mathbf{y})}{P(\mathbf{z})} \tag{9.5}$$

for channel output **z** and input (trial sequence) **y**, where **z** and **y** are usually taken to be branches of the code tree.

One of Fano's contributions was the inclusion of a bias term, giving the *branch metric* or likelihood function for the *i*th branch:

$$\lambda_i = \log \frac{P(z_i|y_i)}{P(z_i)} - B \tag{9.6}$$

As discussed in Section 8.4, the logarithm in (9.6) reduces to a squared difference between the (quantized) ideal and actual demodulator outputs for each channel symbol.

The branch likelihood functions are combined to give a path likelihood function:

$$L_n = \sum_{i=1}^{n} \lambda_i \tag{9.7}$$

the value and behavior of which, for $n = 1, 2, \ldots,$ determine the movement of the decoder through the code tree.

The threshold levels T_i with which values of L_n are compared are taken to be uniformly spaced with increments of T_0. Fano's procedure uses whichever value of T_i is most suitable, as you will soon see.

Assume binary data at the input to the encoder and therefore binary representation of the transmitted and trial sequences throughout the encoding and decoding processes. At the input to the decoder, however, there may be 2^n quantized levels (i.e., soft decisions with n-bit quantization) as a way of preserving and using additional information about the received channel noise. (Recall that $n = 1$ gives the case of hard decisions, i.e., either 0 or 1 as the received channel digit.) Regardless of the number of levels of quantization, it should be noted that the use of a binary code means that each node in the code tree can have exactly two branches coming from it leading to the node corresponding to the next information bit. The effect of the quantization will appear as additional possible values for branch and path likelihoods.

In Fano sequential decoding, it is helpful to make use of the concept of a search tree, which corresponds to (but is not the same as) the code tree. The search tree displays quantitatively, in terms of the values of the various branch and path metrics that result, the movement of the decoder through the code tree along various trial sequences. Thus the search tree is not a redrawing of the code tree. For the metric of (9.6), upward motion in the search tree (that is, increases in L_n) corresponds, in general, to correct hypotheses, regardless of whether a 1 or 0 was sent, For example, a correctly hypothesized bit with low-level channel noise will produce a branch with a terminal node in the search tree that is higher than it would have been in the event of greater received noise because $P(\mathbf{z}_i|\mathbf{y}_i)$ for the closest $\mathbf{y}_i$ will be greater than the *a priori* $P(\mathbf{z}_i)$.

Before discussing the actual operation of the Fano sequential decoding algorithm, one more set of concepts must be introduced. It is necessary to make a distinction between the concept of being at or moving forward or backward to a node, and the concept of looking ahead or back to a node. In each of these cases, the path metric value for the node in question is computed, and a comparison is made between this value and one of the threshold values. The distinction between the two concepts becomes apparent when you consider the node that is serving as a reference point for the current portion of the search for the correct path. Being at (or moving to) a node signifies that a decision has been made to accept for the present the branch and trial sequence terminating at that node. In looking ahead (or back) to a node, there is only a tentative look at the node, with acceptance only a possibility.

The Fano sequential decoding algorithm is shown in simplified form in Figure 9.1, 9.2. The terminology used there has been made as general as possible so that the steps can be applied without rewording to any desired decoding metric. In particular, either Hamming distance or a likelihood ratio can be used for the decoding metric. These two metrics correspond, respectively, to a desire to minimize or maximize the value of the path metric for the correct path. Note, however, that on a short-term (3–4 nodes) basis, the path in the search tree may actually head in a direction opposite to that expected for the correct path, usually as a result of increased channel noise. It is the long-term behavior of the path metric and search tree path that is important.

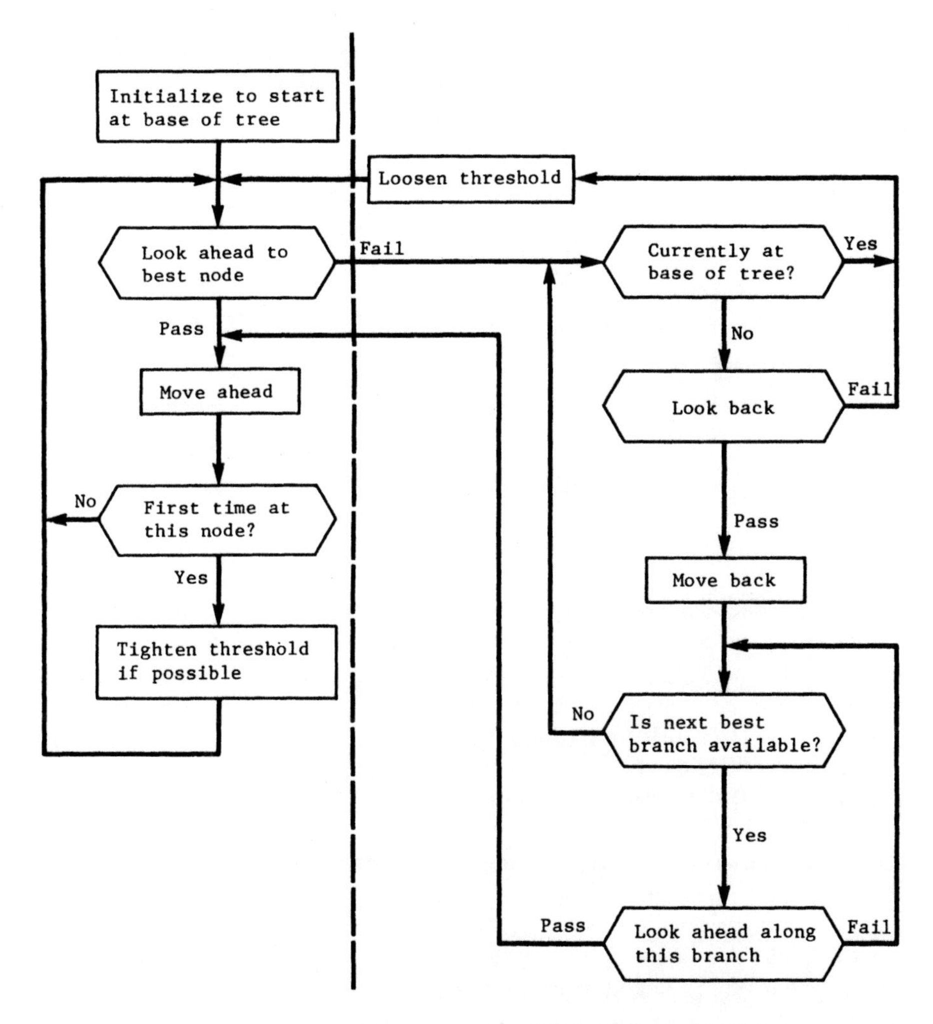

Figure 9.1 Flow diagram of Fano sequential decoding algorithm.

The portion of Figure 9.1 to the left of the dashed line defines the steady-state mode, in which the decoder advances steadily through the search tree. Decoding decisions, when wanted, can be obtained for any digits that are at least one constraint length behind the branch that was just accepted by the box labeled "move ahead" in the forward loop. The portion of Figure 9.1 to the right of the dashed line constitutes the search mode. A *decoding computation* has taken place whenever the decoding metric is computed for a branch. In terms of the flow diagram, this occurs whenever a "look ahead" box is executed.

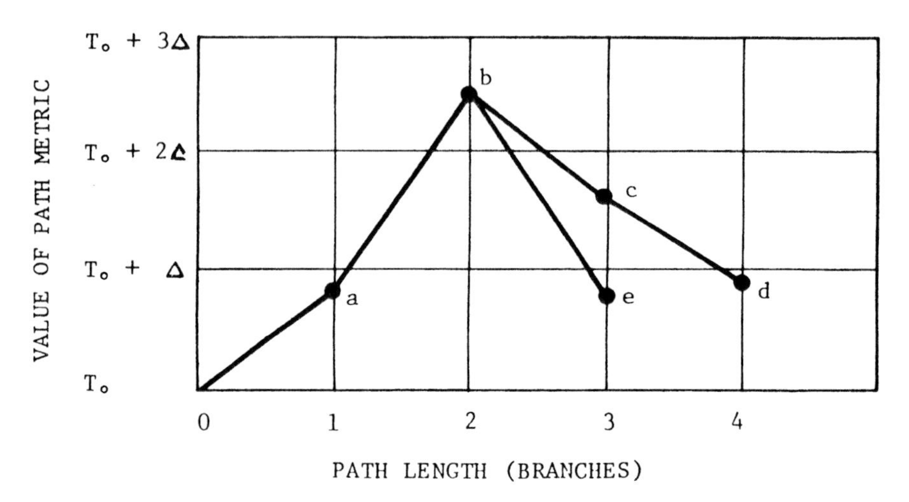

Figure 9.2 Paths in search tree.

The decoder stays in the steady-state mode as long as the sequence resulting from adding the newest branch is accepted, which is the "pass" path from the "look ahead to best node" box. A "fail" decision at this point, however, puts the decoder in the search mode. The first thing the decoder must do is to determine if it is already at the base of the tree. If it is, then since the best branch there did not pass, the decoder cannot advance until the threshold is loosened. This step is repeated if necessary. If the decoder is not at the base of the tree, then it obviously must retreat (look back) to the preceding node in the path. If the value of the path metric there does not satisfy the threshold, the threshold is loosened and the decoder returns to the steady-state mode to attempt to extend the path with this looser threshold, starting at the node from which it just looked back. That is, the decoder does not actually move back if the path metric fails at this point. If, however, the value of the path metric does satisfy the threshold, the decoder then moves back to this node and begins to try other branches if one or more are available. When a branch is found that enables the path metric to pass, the decoder returns to the steady-state mode. If no satisfactory branch is found at this node, the decoder, after checking as to whether it is at the base of the tree, backs up another node.

Example 9.1. The best way to gain an understanding of the operation of the algorithm is to work through an example. In order to maintain a focus on the algorithm itself without digressing to compute branch and path metrics, take the search tree of Figure 9.2 as the starting point. The tree shows only the path values and thresholds. Note that it is irrelevant for the present discussion to know whether a branch at a node corresponds to a 0 or a 1 information digit; therefore no attempt has been made here to identify branches in this way. A metric function has been assumed (such as logarithm of a likelihood ratio) that tends to increase with sequence length; that is, while there may be short-term decreases in the metric value, the overall trend is upward.

In Figure 9.2 the abscissa shows path length, starting at the base of the tree; the ordinate specifies the value of the path metric and shows the various possible threshold values. The column headings of Table 9.2 should be self-explanatory. You should correlate each step of Figure 9.2 and Table 9.2 with movement through the steps in the flow diagram of Figure 9.1.

Some key points about the Fano algorithm can be summarized as follows:

1. While the algorithm could output a decoding decision any time the trial sequence reaches a length of one constraint length and satisfies the threshold, it generally is programmed to defer a decision until at least three to five constraint lengths have been processed successfully. This strategy, while reducing the probability of a decoding error, requires increased storage for received digits and branch metrics plus increased bookkeeping.
2. The Fano algorithm attempts to tighten the threshold after each successful extension of the trial sequence (i.e., regardless of sequence length).
3. The metric includes a bias term the effect of which is to render the various thresholds constant as a functtion of sequence length, making visualization easier during analysis and potentially decreasing storage requirements during implementation.
4. The algorithm must store only one baseline threshold value and one fixed increment value from which all other necessary threshold values are computed. Note that there is no upper bound on the threshold value.

Table 9.2 Moves and rationales for Example 9.1

Path length and node	*Loop: forward (F) or search (S)*	*Looking at node*	*Current threshold*	*Pass (P) or fail (F)*	*Move*	*Threshold increment*
0	*F*	*a*	T_0	*P*	Forward	0
1 (a)	*F*	*b*	T_0	*P*	Forward	$+2\Delta$
2 (b)	*F*	*c*	$T_0 + 2\Delta$	*F*	None	0
2 (b)	*S*	*a*	$T_0 + 2\Delta$	*F*	None	$-\Delta$
2 (b)	*F*	*c*	$T_0 + \Delta$	*P*	Forward	0
3 (c)	*F*	*d*	$T_0 + \Delta$	*F*	None	0
3 (c)	*S*	*b*	$T_0 + \Delta$	*P*	Back	0
2 (b)	*S*	*e*	$T_0 + \Delta$	*F*	None	0
2 (b)	*S*	*a*	$T_0 + \Delta$	*F*	None	$-\Delta$
2 (b)	*F*	*c*	T_0	*P*	Forward	0
3 (c)	*F*	*d*	T_0	*P*	Forward	0
4 (d)			T_0			

9.5 "STACK" DECODING

The following discussion of the so-called stack decoding algorithm is presented to point out the benefits of this simple but powerful technique. Even though no commercially available implementation of it exists, you should be aware of the potential of this technique both for the limits it may impose and for consideration in an advanced development or research situation.

First the advantages offered by the stack algorithm will be presented, followed by a description of its operation relative to that of the Fano algorithm. (The material to be presented here follows Jelinek [9.4].) The chief advantage of the stack algorithm is a greatly reduced average number of computations per decoded digit, relative to this same quantity for the Fano algorithm. This advantage of the stack algorithm has been explored by Jelinek via simulation on a general-purpose computer and is shown graphically in Figure 6 of his paper [9.4]. Results given there show that the stack and Fano algorithms are nearly equivalent at $R = 0.8\ R_{comp}$, with the stack algorithm showing an advantage of about 2:1 at $R = 0.9\ R_{comp}$ and better than 7:1 at $R = R_{comp}$. The stack algorithm has an additional benefit. The Fano decoder can do only the single computation per branch required during quiet intervals of the channel noise, and thus cannot "work ahead" to build an advantage to be drawn upon during noisier periods. In contrast to this, if the highest path in the stack is as long as the received sequence, the stack decoder can compute the path likelihood for the next highest path to full length and store it for potential future use. This last statement correctly suggests that computer memory is the commodity being traded for the impressive speed advantages of the stack decoder. As a side benefit for those interested in following the derivations of results, the stack algorithm is analytically far more tractable than the Fano algorithm. (Compare, for example, the effort leading to the expressions for number of computations in an incorrect subset and for the upper bound on the probability of error in the original works of Reiffen, Fano, Zigongirov, and Jelinek.)

While the stack algorithm is valid for an arbitrary number b branches leaving each node of the code tree, we shall for simplicity develop and explain it in terms of a binary tree; i.e., $b = 2$. The stack algorithm can be described in terms of the following steps:

1. Compute branch likelihood λ for the newest 0- and 1-branches.
2. Compute path likelihoods $L_0(n_1)$ and $L_1(n_1)$ for the paths terminating in the 0-branch and 1-branch, respectively, where n_1 is the number of branches in the path.

3. Arrange $L_0(n_1)$, $L_1(n_1)$ and any other available path likelihood values in descending order. Eliminate the lowest path value in this set. (Note that this ordering and elimination are done without regard to the corresponding path lengths.)
4. For the node corresponding to the largest likelihood value in the stack, compute branch likelihoods and path likelihoods corresponding to each of the two branches leaving that node; then eliminate from the stack memory the likelihood value for the path out to that node and insert in the stack, in proper position according to descending order, the two path likelihoods just found.
5. Repeat Step 4 until the largest path likelihood value in the stack corresponds to a path whose length is equal to the (predetermined) decoding delay.
6. Decode. If the decoding delay is equal to the message length, the entire sequence corresponding to the largest stack entry can be presented as output to the user. On the other hand, if the message length exceeds the decoding delay, only the initial branch need be released. In this case, all stack entries corresponding to paths of the incorrect subset (i.e., those paths beginning with the other branch from the base node of the tree) must be purged from the stack before continuing. At the same time, the decoded branch likelihood can be subtracted from the path likelihoods remaining in the stack, although this is not necessary.

Example 9.2. This illustration of the stack algorithm elaborates on the details of the example used by Jelinek [9.4]. The tree to be considered is shown in Figure 9.3. Nodes are numbered for identification and easy reference, and the likelihood value found for each branch is written just above the branch. For this example, it is not necessary to know the correct (transmitted) sequence.

The contents of the stack are shown in Figure 9.4. At each node, the path likelihood L and path length (in branches) are given. The algorithm terminates when a node with path length $= 3$ appears at the top of the stack. The lines at the right of the figure indicate the disposition of the node which was at the top of the stack at the start of that step. At Step 7, the top stack entry corresponds for the first time to a path which is three branches long. Through stored information giving linkages between nodes, the decoder determines that path 8 corresponds to an initial information bit value of 0. It decodes this first bit (or the entire path through nodes 1, 3, and 8, depending on circumstances).

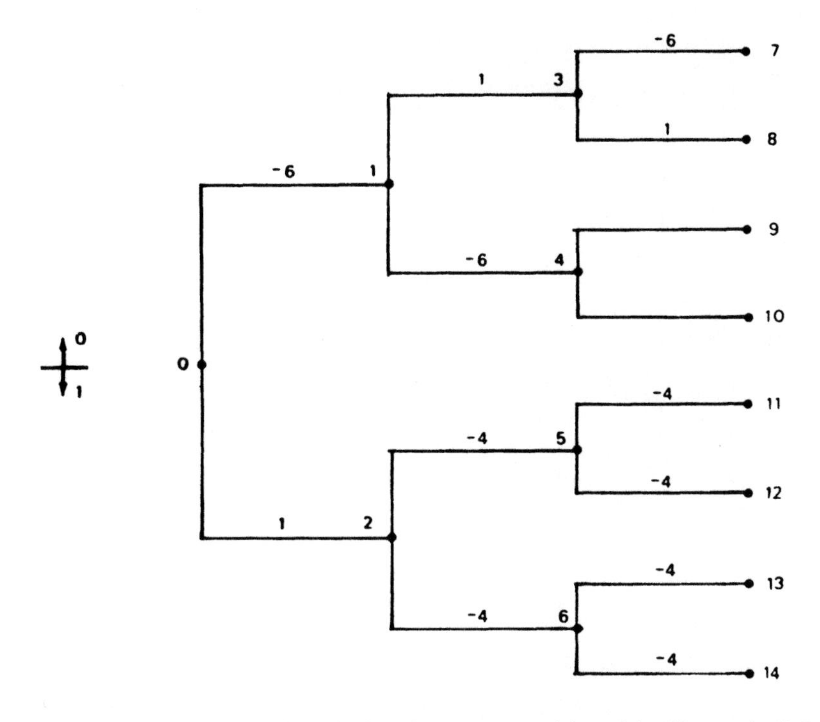

Figure 9.3 Branch likelihoods for the tree considered in Example 9.2.

Following this decoding decision, the likelihood of the decoded branch can be subtracted from the values of L for nodes in the correct subset (in this case nodes 8, 7, and 4). Nodes in the incorrect subset (in this case 11, 12, 13, and 14) can be purged from the stack. This leaves the values of Table 9.3.

Table 9.3

Node number	L	*Path length*
8	2	2
7	−5	2
4	−6	1

Step	Node Number	L	Path Length
1	0	0	0
2	2	1*	1
	1	-6	1
3	5	-3*	2
	6	-3	2
	1	-6	1
4	6	-3*	2
	1	-6	1
	11	-7	3
	12	-7	3
5	1	-6*	1
	11	-7	3
	12	-7	3
	13	-7	3
	14	-7	3
6	3	-5*	2
	11	-7	3
	12	-7	3
	13	-7	3
	14	-7	3
	4	-12	2
7	8	-4**	3
	11	-7	3
	12	-7	3
	13	-7	3
	14	-7	3
	7	-11	3
	4	-12	2

*Largest L in stack, but not for full-length sequence.
**First time largest L achieved for full-length sequence.

Figure 9.4 Example of stack algorithm operation.

Jelinek has compared his version of the stack algorithm with the Fano algorithm on the basis of the required average decoding speed advantage $\overline{A}$ (defined as the number of decoding steps performed in the time one branch is received). A code having $R = 1/2$ was used on the BSC at various crossover probabilities p up to $R_{comp} = R$, which corresponds to $p = 0.045$. The algorithms are virtually identical at $R/R_{comp} = 0.8$ with $\overline{A} = 1$. At $R/R_{comp} = 0.9$, $\overline{A}$ is about 2 for the Fano algorithm and still about 1 for the stack algorithm. At $R = R_{comp}$, these values are, respectively, 14 and 2, showing the significant speed advantage of the stack algorithm at high information rates for the BSC.

9.6 DECODER PARAMETERS

The operation of two variations of sequential decoding has been described. It is clear that there will usually be a great variability, as a function of time, in the number of decoder computations necessary to extend the trial sequence by one branch. This translates into a similar variability in the number of computations required to decode one digit. The effect of this variability on decoder hardware, and eventually on availability of received data, is profound. Long searches, which result when a particularly noisy received sequence is encountered, cause the continuing stream of incoming digits to accumulate in the meantime. In order to preclude the loss of these digits, there must be an input buffer to store them. To minimize the number of digits that enter the buffer during a search, the decoder must have a speed advantage (or speed factor, as it is often called) on the order of ten times the information rate. Nonetheless, this buffer will, with probability one, overflow at some time. Thus the user must be prepared to request retransmission or to lose some data.

As pointed out in Section 9.2, the decoder computation effort depends strongly on the ratio R/R_{comp}. In particular, there is the following result, due to Gallager [3.1]: For sequential decoding applied to a DMC, the average number of forward moves (in the search tree) per decoded subblock of $k_0 = n_0 R$ information digits is bounded by

$$4/[1 - e^{n_0(R - R_{comp})}], \; R < R_{comp} \tag{9.8}$$

Clearly, this expression becomes infinite as $R \rightarrow R_{comp}$.

Typical values of code constraint length $n_A = n_0 K$ for sequential decoding range from about forty up to several hundred channel digits. High information rates ($\geq 1/2$ bit/symbol) and channel rates (on the order of tens of megabits per second) have been achieved in commercially available hardware offered by several vendors.

9.7 SUMMARY AND CONCLUSIONS

Both the power and the limitations of sequential decoding have been described. The power of this technique arises from its ability to provide bit error rates that decrease exponentially with code constraint length at information rates below channel capacity, in combination with a computation load with moments of all orders that are bounded by a constant independent of constraint length as long as $R < R_{comp}$. Furthermore, storage requirements grow linearly with code length.

The limitation is the variability in computation time to advance one node or branch. This variability forces the decoder to include buffer storage amounting to some small number of constraint lengths and to possess a speed advantage, relative to the channel data rate, to enable it to keep up during noisy periods or at least to be able to catch up quickly once the channel has again become quiet. Because of this behavior on noisy channels, sequential decoding, like Viterbi decoding, is unsuitable as a burst-correcting technique without additional help such as interleaving (Chapter 10).

Chapter 10
Coding to Combat Impulse Noise

10.1 INTRODUCTION

This chapter discusses techniques of coping with bursts of errors, as opposed to the random errors for which most schemes are designed. We shall consider some codes such as the Reed-Solomon codes previously discussed, along with other techniques. As you may be aware, in considering real-life communication channels, one frequently encounters channels the behavior of which is strongly influenced by impulsive noise having large amplitude and relatively short duration. At high transmission rates, even a short noise pulse can affect a large number of bits. The effect of such noise is the production of a burst of errors in the received channel symbols. This chapter looks at ways of coping with error bursts in the case of binary transmission and considers techniques both with and without a feedback channel, with emphasis on the latter. In that case the burden of error control falls squarely on forward error correction by means of either block or convolutional codes. The material of this chapter applies and, for threshold decoding and code combining, extends results obtained in earlier chapters.

We begin with a definition: a noise *burst* of length b is usually defined relative to a *guard space* of length g. The guard space contains all 0s as error digits. The burst is a sequence of b binary digits such that

1. the first and last of its digits are 1s (errors);
2. a guard space must precede and follow the noise burst; and
3. the sequence of b burst digits does not include a sequence of g or more consecutive 0s.

Note from this definition that not all of the burst need be 1s.

Example 10.1. The noise sequence 00001100111000 could be regarded as containing a burst of length 7 (fifth through eleventh digits) relative to a guard space of length 3, or it could be considered as containing bursts of lengths 2 and 3 relative to a guard space of length 2.

Example 10.2. The sequence 0001001000 can be viewed either as containing two random errors or a burst of length 4 relative to a guard space of length 3.

10.2 CODING WITH FEEDBACK

The simplest scheme for handling bursts consists of error detection and retransmission using block codes (for a full and detailed discussion, see Lin and Costello [5.8] Chapter 15). In this scheme, an easily implemented (e.g., cyclic) code is usually used. A block diagram of a reasonably sophisticated system is shown in Figure 10.1. The noiseless feedback channel is attainable in practice because of the very low data rate of the feedback signal, which permits large values of E_B/N_0. Before considering the operation of this system, note that an elementary version of it might be one in which the transmitter waits after sending out a code word until it receives a signal that the word was either correctly received (ACK) in which case it sends the next word, or was incorrectly received (NAK) in which case it retransmits the word just sent.

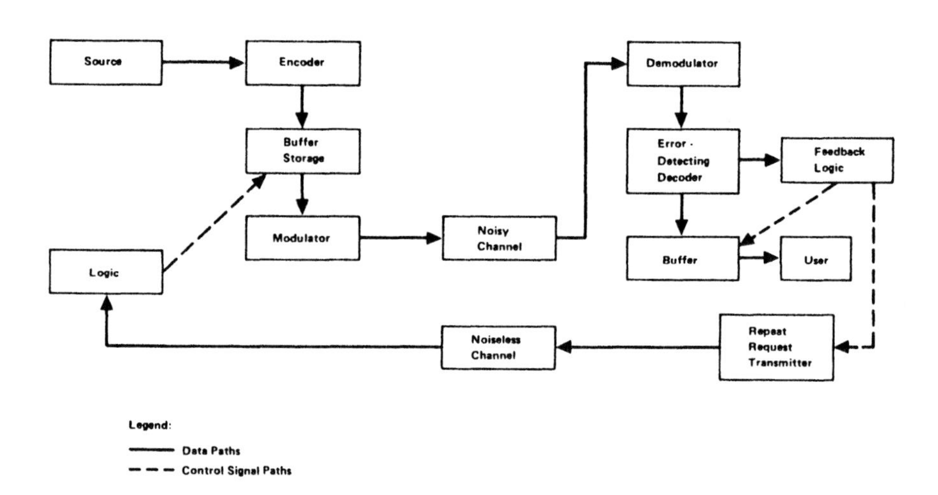

Figure 10.1 Error control using error detection and retransmission.

At the receiver, the syndrome **s** is computed. If $\mathbf{s} \neq \mathbf{0}$, one or more errors has occurred and a request for retransmission is sent; if $\mathbf{s} = \mathbf{0}$, the code word is sent on to the user under the assumption that no errors have occurred. In the situation shown in Figure 10.1, storage beyond that necessary for error correction is required at both the receiver and the transmitter. As with error correction, the receiver must store the word following the one being decoded. In addition, if the decoding results in the detection of one or more errors, the decoder initiates a repeat request, by which time the second word following the word just decoded is on its way. If either or both of these last two words turn out to be error-free, they can be passed on to the user immediately, once their turn comes to be decoded. The situation at the decoder forces the same three words to be stored at the transmitter, in addition to subsequent words if the data source cannot be turned off at will. Logic must be provided at the receiver to initiate a repeat request when necessary and to enable the buffer to release information bits from a correct codeword to the user. Logic at the transmitter responds to a repeat request by signaling the buffer there to retransmit the proper word. This logic would also turn off the source during periods of retransmission if that happens to be possible.

What is the performance capability of the system just described? Two performance measures are important: the probability of an undetected error and the throughput rate.

An undetected error occurs when $\mathbf{s} = \mathbf{0}$ because the received error sequence was actually a code word. Since an (n, k) binary code contains 2^k code words each of which could be received as any of 2^n possible sequences, the probability of receiving such a sequence is very nearly $2^k/2^n$ or $2^{-(n-k)}$. Thus, for example, for the modest values $n = 31$, $k = 16$, this probability would be 2^{-15} or about 3×10^{-5}, a value that would be more than adequate for many purposes. (The exact value of probability of undetected error can be found from

$$P_u = \sum_{i=0}^{n} A_i p^i (1 - p)^{n-i}$$

where A_i is the number of code words of weight i and p is the channel bit error probability.)

The following theorem relates code redundancy to the ability of the code to detect a single burst of a specified length.

Theorem 10.1. For detecting any burst of length b or less with a linear code of length n, b parity-check symbols are necessary and sufficient.

Thus, in particular, an $(n, n - b)$ cyclic code will detect up to b consecutive errors occurring anywhere within a code word. For the example in the preceding paragraph, $b = 15$. It is also useful to note that the minimum distance $2t + 1$ required to correct all patterns of t errors is adequate to *detect* all patterns of $2t$ (not necessarily consecutive) errors. Conversely, if only a t-error-detection capability is desired, then a minimum distance of $t + 1$ is sufficient.

In addition to the probability of undetected error, an equally important measure of effectiveness for schemes involving retransmission is the *throughput rate*, which is the average time rate at which information is accepted at the decoder. Thus, if

N_t = total number of attempts at transmitting code words
N_a = total number of code words accepted at decoder
N_r = total number of repeat requests
R_{tr} = transmission rate in b/s
R_{th} = throughput rate in b/s

then

$$N_t = N_a + N_r$$

and

$$R_{th} = \frac{N_a}{N_t} R_{tr}$$

The scheme described in this section works very well when the forward channel has relatively long intervals of many block lengths of error-free behavior. On the other hand, a prolonged noisy period could result in drastically reduced throughput at best and loss of data at the worst if the input to the encoder cannot all be stored or temporarily halted.

A very thorough investigation, which considers throughput rate as more important than block error rate, has been reported by Benice and Frey [10.1]. They evaluated BCH codes of lengths 7, 15, . . . , and 1023 by means of computer simulation. Similar results are reported in the excellent paper by Burton and Sullivan [10.2].

A most impressive use of BCH codes in an error detection role with feedback was first simulated by Fontaine [10.3] and then implemented by

Reiffen, Yudkin, and Schmidt [10.4]. The repeat request logic used is a refinement of the scheme presented in Figure 10.1. A (255, 231) binary BCH code was used to detect errors on a telephone channel. The predicted time to first undetected error was close to 300 years!

10.3 INTERLEAVING

With interleaving, as shown in Figure 10.2, the data stream (the symbols i_j) is separated into m parallel streams, each of which is encoded separately. The encoded sequences (the symbols x_l) are then commutated for serial transmission over the channel. At the receiver the stream is again split, each resulting stream is decoded, and the decoder outputs deinterleaved to give the original data. This technique can be applied equally well to block and convolutional coding schemes. Its success is based on the fact that for m sufficiently large, bursts of channel errors are spread out over the m separately encoded streams, resulting in the need for much less potent error correction on any given stream. That is, each of the m streams appears to have passed through a memoryless channel.

As shown in Figure 10.2, implementation can be carried out in terms of m encoders. Similarly, m decoders are necessary. If a cyclic code is being used, each encoder and decoder consists of shift registers, mod-2 adders, and some decision logic for the decoders.

A second method is to interleave the stages and taps of the encoder shift registers into a single shift register m times as long. For cyclic block codes, this converts an (n, k) code with generator polynomial $g(x)$ in each of the original m encoders into an (mn, mk) code having generator $g(x^m)$. A similar relationship holds for convolutional codes, in which $m - 1$ stages are inserted after each stage in one of the original m encoders. It can be shown that if the original (n, k) cyclic code had the ability to correct bursts up to b symbols in length, then the (mn, mk) interleaved code can correct bursts up to mb symbols long.

You should be aware that as far as a cyclic code is concerned, a burst of any length b_1 can just as well occupy the first $b_2(< b_1)$ positions and the last $b_1 - b_2$ positions of the codeword; i.e., it can be an "end-around" burst.

Finally, for block codes or for a convolutionally encoded block of data, there is the technique shown conceptually in Figure 10.3. With a block code of length n and interleaving depth m, each word is read into a row of the storage matrix (after encoding). The bits are then read out

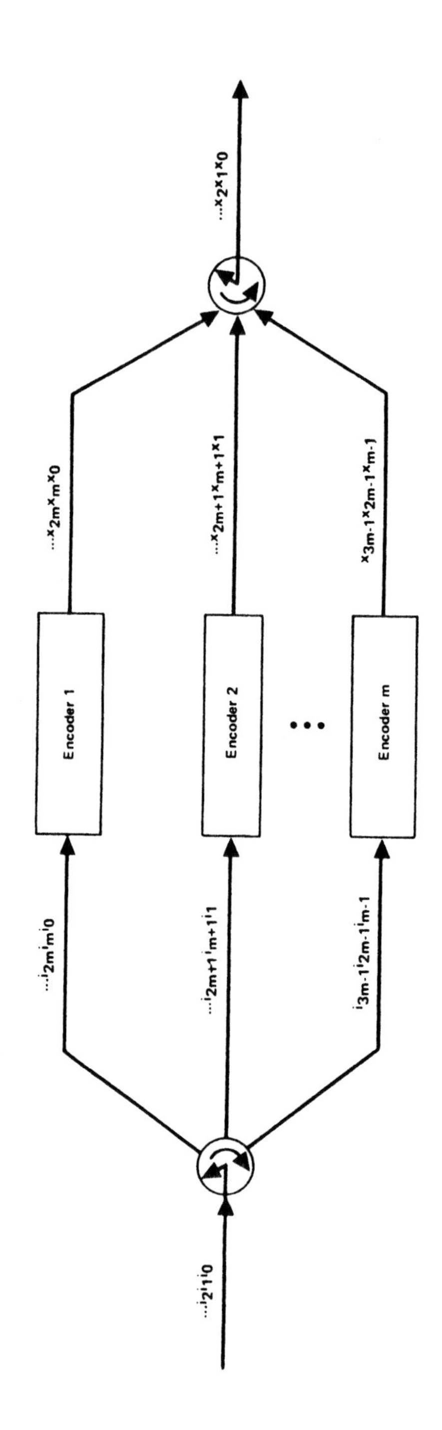

Figure 10.2 Interleaved encoding.

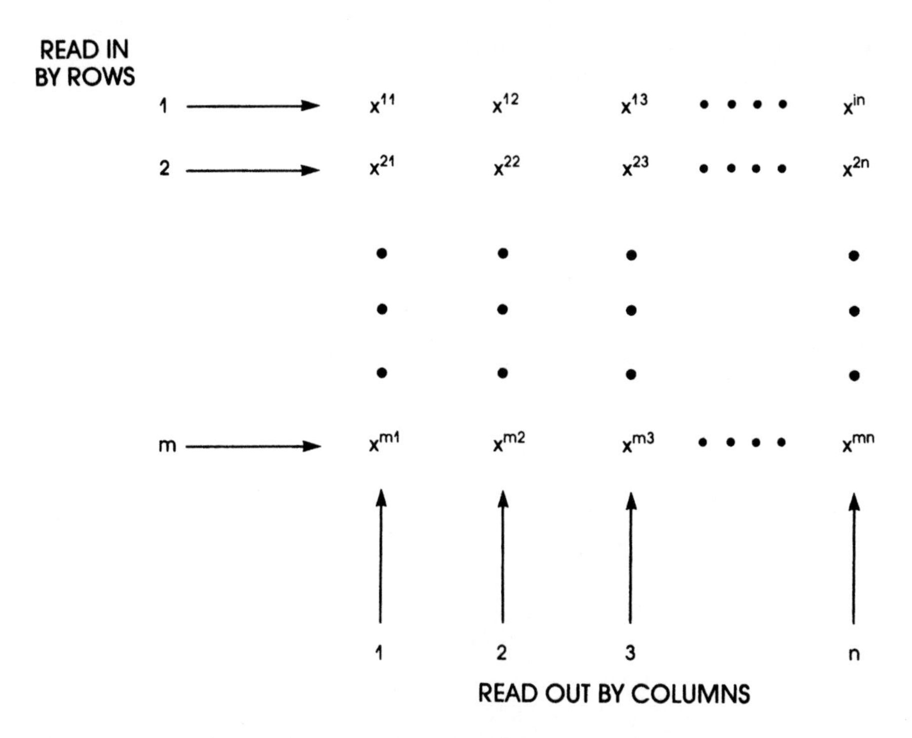

Figure 10.3 Storage matrix for interleaving.

by columns. At the receiving end this procedure is reversed. For convolutionally encoded data, where there is no block structure, the first n encoded channel bits are read into the first row, the next n into the second row, *et cetera*. If the total length of the transmitted data is less than mn, filler bits can be added at the end. The readout into the modulator is by columns, just as for block codes.

10.4 BURST-CORRECTING BLOCK CODES

In contrast to interleaving are techniques that rely on codes especially constructed to deal with bursts of channel noise. The following theorem, which holds for nonbinary as well as binary codes, is useful in evaluating the capabilities and efficiencies of these codes.

Theorem 10.2. In order to correct any possible error burst of length b or less, a linear code must have at least $2b$ parity-check symbols. In order to correct any burst of length b or less and simultaneously detect any burst of length $d \geqslant b$ or less, a linear code must have at least $b + d$ parity-check symbols.

The two best known classes of burst-correcting block codes are the Fire [10.5] and Reed-Solomon [5.14] codes. The latter were described in Chapter 5, and the former are discussed below. (Peterson and Weldon [3.2] present a much more thorough discussion of burst-correcting codes, including a list of good codes.) The unpleasant truth is that for both block and convolutional burst-correcting codes, code design is still largely an art, leaning heavily on trial-and-error techniques. Encoding for the Fire and RS codes is done just as it is for any other cyclic code, with arithmetic in GF(2^m) for the RS code.

A general burst-correcting decoder is shown in Figure 10.4. As usual, the values of g_i are the coefficients in generator polynomial $g(x)$ and are therefore implemented as open circuits or wires for $g(x)$ over GF(2). This decoder contains only delay elements, mod-2 adders, gates, and combinational logic.

As with all cyclic codes, a Fire code is most easily defined in terms of its generator polynomial, which is of the form $g(x) = p(x)(x^i + 1)$, over GF(2). Here $p(x)$ is an irreducible polynomial of degree m with roots of order e, and $p(x)$ and $x^i + 1$ are relatively prime. Then code length n is the least common multiple of i and e. The number of parity checks is $i + m$, so that the number of information symbols is $k = n - i - m$. A Fire code can correct a single burst of length b or less and simultaneously detect a burst of length $d \geqslant b$ or less if $i \geqslant b + d - 1$ and $m \geqslant b$.

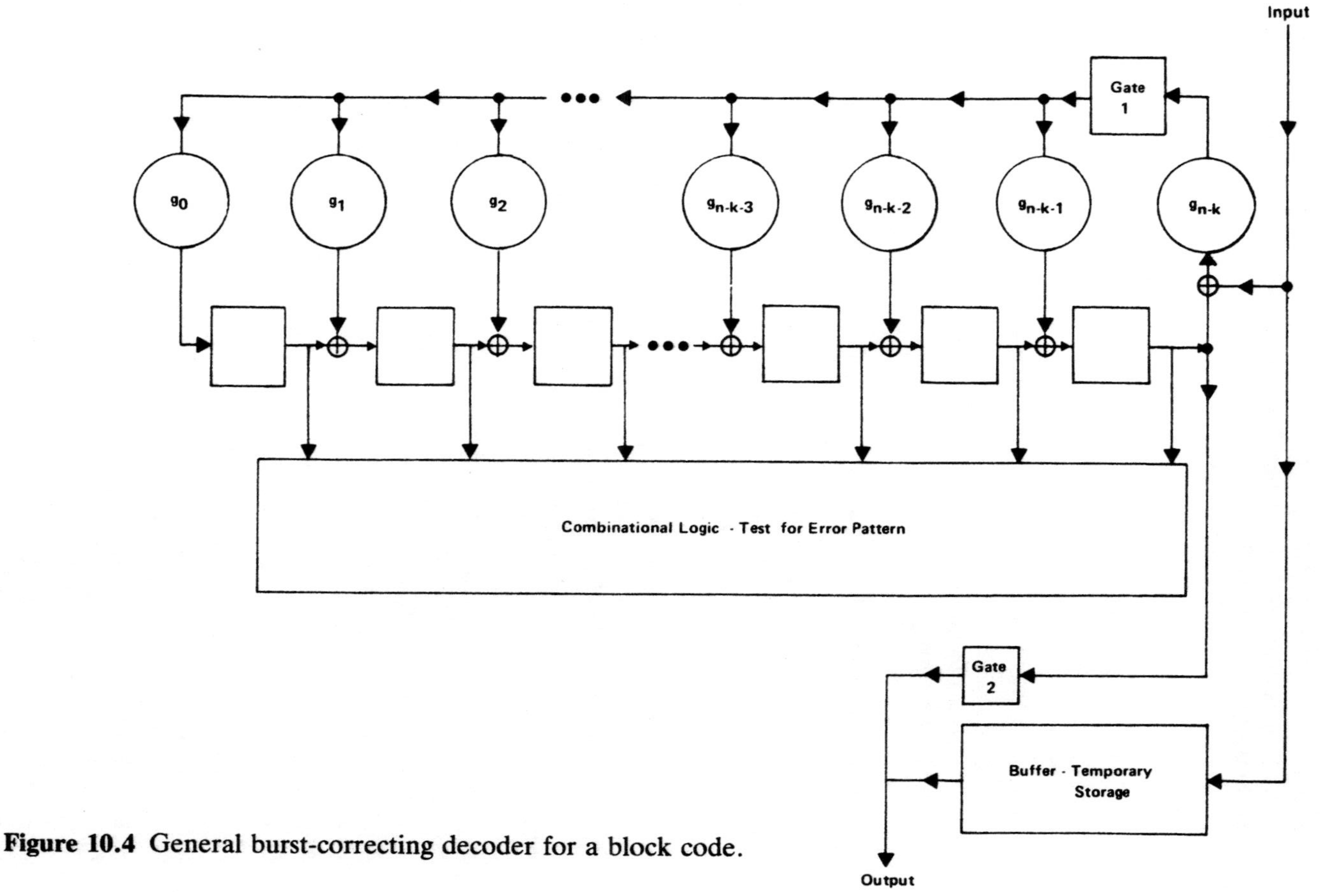

Figure 10.4 General burst-correcting decoder for a block code.

Alternatively, it can detect any combination of two bursts in which the length of the shorter burst is no greater than m, and the sum of the lengths is no more than $i + 1$; or it can certainly detect any single burst of length $i + m$. The factor x^{i+1} gives rise to pairs of parity checks spaced i digits apart, so that no burst of length less than i will affect both checks of a pair.

First, compare the ability of a Fire code to detect or correct bursts with the performance bounds of Theorems 10.1 and 10.2. According to Theorem 10.1, $i + m$ parity checks are necessary and sufficient to detect a burst of length $i + m$, which is exactly what is claimed for the Fire codes. Applying the bounds of Theorem 10.2 to the simultaneous burst correcting and detecting capability, you can see that the number of parity check symbols must satisfy $i + m \geq b + d$, which is certainly true for the nontrivial case $b > 1$, since $i \geq b + d - 1$ and $m \geq b$.

Finally, note that since the number of parity checks is $i + m$, this fact combined with the relationships $m \geq b$, $i \geq b + d - 1$, and $d \geq b$ implies that $i + m \geq 2b + d - 1 \geq 3b - 1$, which is well in excess of the minimum number required to detect or correct a burst of length b.

Example 10.3. Let

$$g(x) = (x^7 + x^2 + 1)\,(x^{16} + 1)$$

With $x^{16} + 1 = (x + 1)^{16}$, you can verify that $x + 1$ does not divide $x^7 + x^2 + 1$ evenly. Then $m = 7$, $i = 16$, and it turns out that $e = 127$, which is prime. Then $n = ie = 2032$, $n - k = 7 + 16 = 23$ parity checks, and $k = 2032 - 23 = 2009$ information digits. You must therefore choose $b \leq m = 7$. Choosing $b = 7$, find $10 \geq 7 + d - 1$. That is, the code corrects any burst of length 7 or less and simultaneously detects a burst of length 10 or less; if used for pure detection, it can detect a single burst of length 23 or less. In this example, then, you have a code in which a code word of more than 2000 symbols is required to correct bursts no more than 7 symbols long. On the positive side, the rate in transmitted symbols per bit of information is high, nearly 0.99; the implementation is extremely simple, requiring only $n - k = 23$ stages in the encoder shift register.

The Fire codes are not as efficient as some more recent additions to the set of known burst-correcting codes. For example, Burton [10.6] has defined codes of length $n = be$ which correct any burst of length $b_1 \leq b$, as long as such a burst starts at the 1st, $(b + 1)$th, $(2b + 1)$th, . . . , or $(eb + 1)$th position of the code word. These codes have

$$g(x) = p(x)\,(x^b + 1)$$

with $p(x)$ and $x^b + 1$ relatively prime and $p(x)$ of degree b with roots of order e.

This raises the question: when can a Fire code be used profitably? The answer is simple: if you want protection against short to fairly long bursts with a high-rate, easily implemented code, then use a Fire code or one by Burton or any of several others (see, for example, Table 11.1 in Peterson and Weldon [3.2]). Such a code is ideally suited to detecting errors occurring in data stored in a digital computer.

If a higher-order code alphabet is used, more powerful burst-correcting codes are possible. Thus, if you group m binary digits together, each symbol thus formed is an element of an alphabet of $M = 2^m$ symbols. With this point of view, you see that a burst of length $(t - 1)\, m + 1$ binary digits or less can affect at most t successive symbols from the M-ary alphabet. This is the basis for the use of Reed-Solomon (RS) codes, discussed in Chapter 5, in burst correction. By Theorem 10.2 an RS code that corrects t errors in an M-ary code word needs at most $2t$ M-ary check symbols (i.e., $2tm$ binary check digits) out of the $n = 2^m - 1$ M-ary symbols in a code word. Note that this makes it possible to correct more than one binary burst contained within the M-ary burst of length t.

Example 10.4. If $m = 7$ and $t = 4$, the RS code has symbols consisting of blocks of 7 binary digits each. The code can correct all bursts of length $(t - 1)\, m + 1 = 22$ binary digits or less using $2tm = 56$ binary check digits out of a word length of $m(2^m - 1) = 889$ binary digits. This compares very favorably, from the standpoint of code efficiency and burst-correcting power, with the Fire code discussed previously. Decoding is slightly more complicated, requiring a special-purpose computer to carry out finite-field algebra. Advances in theory and technology have resulted in substantial reductions in decoding time and complexity. It is of passing interest to observe that a Fire code to correct a burst of length 22 or less requires only 65 binary parity-check symbols but has a code length of possibly 1.6×10^8 binary digits.

10.5 BURST CORRECTION WITH CONVOLUTIONAL CODES

This section briefly discusses two techniques (there are others) that can be applied to correct bursts when using convolutional codes. Both techniques relate to threshold decoding, and hardware may be commercially available for implementing both. They are characterized by the ability to cope with very large bursts with little increase in equipment complexity as they require only additional delay units in direct proportion to the increase in burst length b.

The first technique is diffuse threshold decoding, developed in the early 1960s and later published by Kohlenberg and Forney [10.7]: The encoder and decoder are shown in Figures 10.5 and 10.6, respectively. The code illustrated corrects bursts of $b = 2\beta$ binary digits relative to a guard space of $g = 6\beta + 2$. Note that the technique is not quite the same as the interleaving defined earlier, since not all taps of the shift register are separated by exactly β stages.

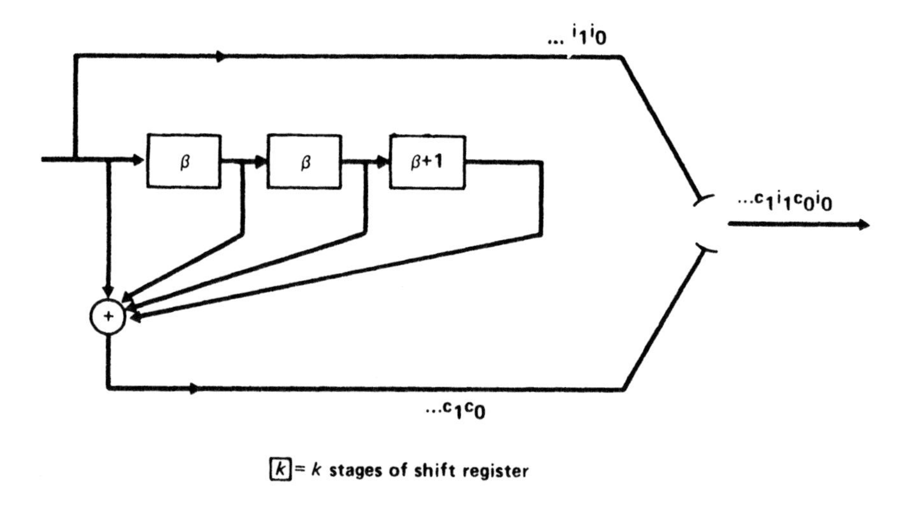

Figure 10.5 Encoder for diffuse threshold decoding.

The second technique is due to Gallager [3.1] and is shown implemented in Figures 10.7 and 10.8. For the code shown, most burst up to $b = 2\beta$ relative to $g = 2(\beta + k_1 + k_2) - 1$ are corrected. Note that $b \approx g$ for $\beta \gg k_1$ or k_2 (typically, $\beta \approx 1000$, while k_1 and k_2 are less than 10). As noted, this power is achieved at the expense of not being able to correct all bursts relative to the specified guard space. Implementation is again extremely simple. Basically, the decoder operates as follows. A burst of errors in the encoder section causes a corresponding burst of 1s to enter the syndrome register. The latter burst is in the last $5 + k_1 + k_2$ stages of this register, and therefore causes an output from the burst detector, at just the time that the leading edge of the channel error burst reaches the right-hand end of the $(\beta + k_1 + 6)$-stage portion of the encoder. With very high probability, this output is the correct estimate of the error in information digit $\hat{\imath}$ about to be emitted from the decoder.

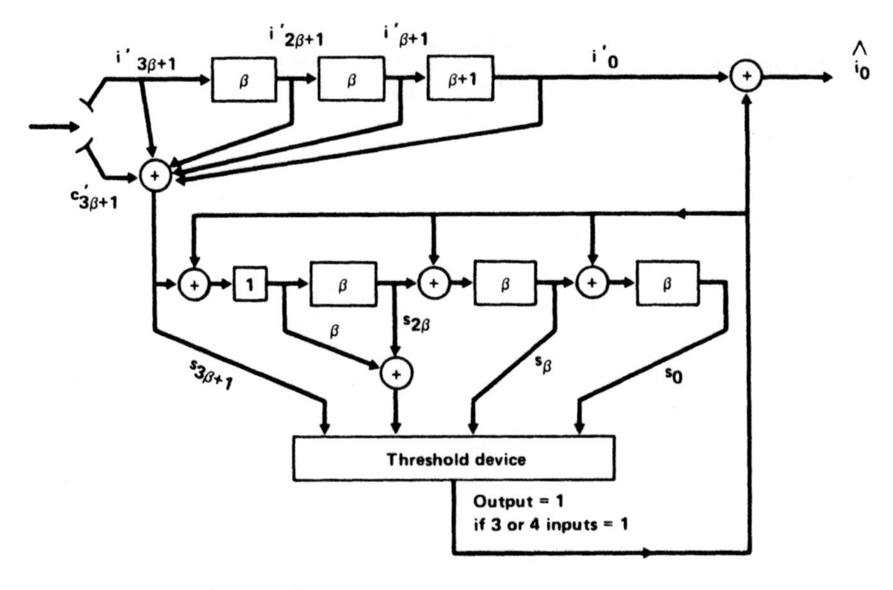

Figure 10.6 Diffuse threshold decoder.

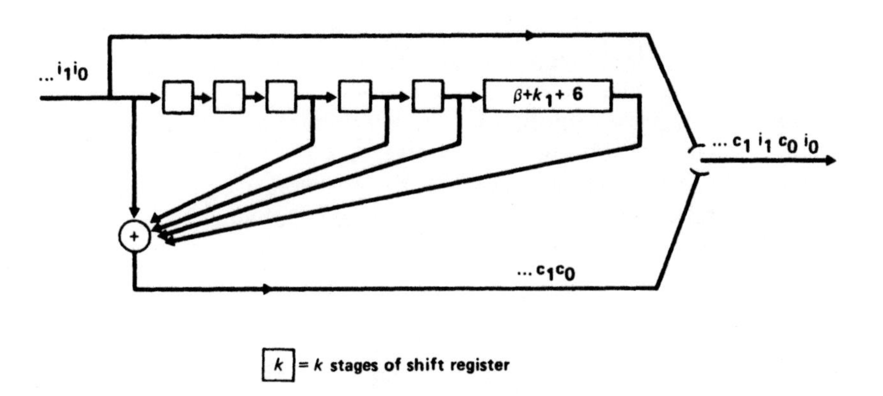

Figure 10.7 Gallager burst encoder.

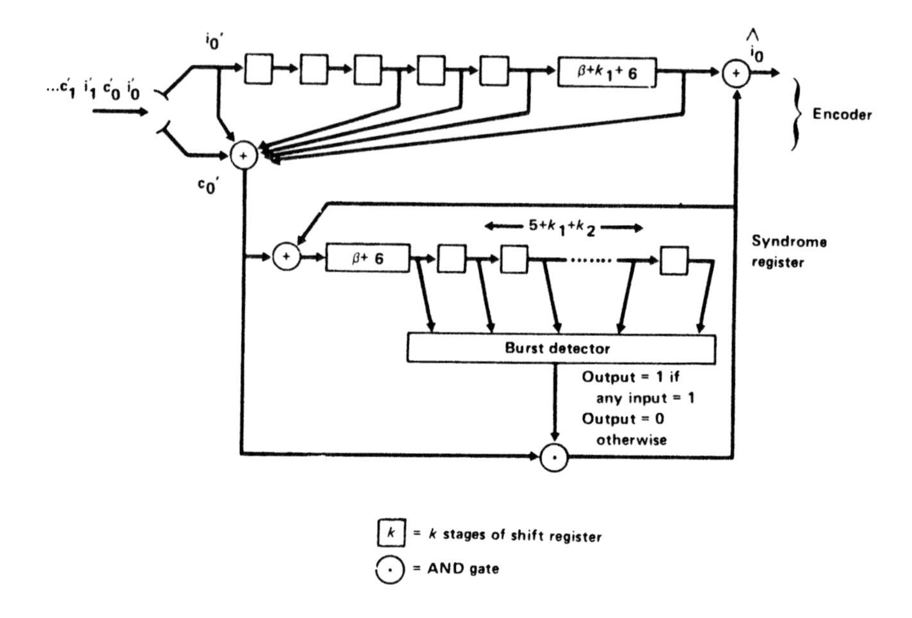

Figure 10.8 Gallager burst decoder.

10.6 CODE COMBINING

A technique, called *code combining,* can adaptively adjust the code rate to match channel conditions as a means of reliable communication even if the average channel bit error probability approaches one-half. This technique accomplishes information transfer under adverse conditions by weighting each received copy of a message. The copy is then combined on a bit-by-bit basis with the sum of all previously received weighted copies of the same message until a composite message is obtained that passes all of the applied error-checking criteria. (An excellent introduction is given in the paper by Chase [10.8].)

The message is usually assumed to be in the form of a packet, with the information bits (and possibly other portions such as a header) being used to compute a cyclic redundancy check (CRC) using one of the standard CRC polynominals. The bits checked by CRC and the CRC bits themselves are then convolutionally encoded and transmitted over the channel. At the receiver, the convolutionally encoded portion is decoded (e.g., with Viterbi or sequential decoding), and then the CRC is applied to the decoded data bits. If the latter check detects no errors, the packet is assumed to have been correctly received and is passed to the user.

However, if the CRC fails, a second copy of the packet is combined with the first and the entire decoding process is repeated on the combined packet. The process is continued until either the CRC test is successful, or some predetermined number of packet repetitions is reached.

Packet weighting is usually set up to reflect average channel conditions. One possibility is to use $w_i = \log[(1 - p_i)/p_i]$ as the weight of the ith packet, where p_i is the estimated average channel bit error probability for that packet. Note that $w_i = 0$ for $p_i = 1/2$ and that w_i becomes very large as $p_i \rightarrow 0$ so that good packet copies are weighted much more than bad copies are. One way to estimate p_i is to include a block of unencoded bits (frequently chosen to be all 0s) with the packet and observe the number of these known bits received in error. Thus, individual encoded bits can be demodulated using hard decisions, while the overall packet weighting effectively makes a soft decision on each copy of the packet. The system described here is shown as a block diagram in Figure 10.9.

Note that the CRC yields the final decision to accept or reject the packet. Generally, the CRC would be one of the current data communication standards (i.e., CRC-16, CCITT-16, CRC-32). Recall that an undetected error can occur only when a code vector (packet) is received as another code vector, and that the probability of undetected error is very closely approximated by 2^{-r} where r is the number of parity-check bits. Thus, CRC-32 gives a probability of undetected packet error of about 2^{-32} or 2.3×10^{-10}. While such a low probability virtually assures that no incorrect packet will be accepted, this high degree of message reliability is obtained at the price of decreased throughput (and therefore, increased delay) when the channel is poor because of interference or fading. There is an obvious trade-off here, and the choice is not necessarily easy to make from the standpoint of user satisfaction. At the one extreme, each packet is repeated as often as necessary to satisfy the CRC. (Bear in mind that a very noisy channel might actually cause even more errors in the convolutional decoder output than there were in the input.) At the other extreme, setting some arbitrary number of repetitions of the packet (e.g., 5 or 6) will certainly reduce delays, but may result in an unacceptably high number of packets being either rejected or delivered with errors to the end user.

Note that with the "repeat until perfect" strategy, this technique actually implements the adjustment of code rate to match channel capacity and comes arbitrarily close to realizing the promise of Shannon's noisy channel coding theorem without wasting channel capacity with a code that is too powerful for channel conditions.

Recently, Kallel and Haccoun [10.9] analyzed code combining using sequential rather than Viterbi decoding of the convolutional code. Their analysis showed that at rates $R > R_{comp}$, code combining is superior to ordinary automatic repeat request (ARQ) but at $R \leq R_{comp}$ there is no improvement over ordinary ARQ.

10.7 SUMMARY

The techniques discussed in the preceding sections for burst noise channels are summarized in Table 10.1. This set of techniques is not exhaustive, but it is certainly representative of those currently available.

Table 10.1 Comparison of burst-correcting techniques

Technique	*Advantages*	*Disadvantages*
1. Error detection and retransmission	Easy to implement. Less redundancy required than to correct same number of errors. Independent of burst length.	Requires reliable feedback link, possibly buffer storage & logic at receiver and transmitter. Low throughput with noisy forward channel.
2. Simple interleaving	"Dilutes" effects of noise burst by distributing over separately coded data streams. Easy to implement; many good random-error-correcting codes are known.	Dependence of errors in burst not fully exploited in decoding. May still be inadequate with very long bursts or fadings.
3. Fire codes	Very high code efficiency. Very simple implementation.	Long codes required for modest burst-correcting capability.
4. Reed-Solomon codes	Code efficiency typically greater than 90%. Code construction easy. Hardware is commercially available.	No serious disadvantages.
5. Threshold decoding techniques	Capable of handling large bursts (~2000). Hardware commercially available.	Code design for bursts is still an art. Equipment may exist for only limited range of code rates.
6. Code combining	Arbitrarily low output error rate, even under most severe conditions. Code rate adapts to channel conditions.	Very low throughput under severe conditions, moderately low under benign conditions, at kb/s channel rates.

Chapter 11
Applications and Trade-Off Analyses

11.1 INTRODUCTION

This chapter presents examples of the application of error-control techniques to situations that occur in practice. Summaries of the trade-offs will be presented. Remember that a technology-based factor such as speed, cost, weight, or power, which was decisive in reaching a certain conclusion at one time, may with new technology be irrelevant or may lead to a different conclusion.

Parameters to consider include:

1. Data rate in bits per second; i.e., the rate at which data are generated and/or must be transferred by the system.
2. Available or required ratio of signal energy per bit to noise power density (E_B/N_0).
3. Communication channel or storage medium characteristics—Gaussian, impulse, or other type of noise; depth, duration, and frequency of occurrence of fades, if any; interference, either accidental (e.g., other users) or intentional (jamming).
4. Availability of processing (encoding and decoding) and implementation (software *versus* hardware, general- *versus* special-purpose computer).
5. Processing speed compared to data rate; possible need for buffering to allow for variations in rate at which processing can take place (e.g., input buffer with sequential decoding).
6. Tolerances of the message (data, digital voice, video) to errors; i.e., required output bit or block error rate. (Video may still be acceptable at $P_B = 10^{-1}$, voice at 10^{-3}, while data may require 10^{-4} down to 10^{-9}.)
7. Required throughput rate.
8. Existence and characteristics of a feedback link.

9. In case of jamming, expected type, duty cycle, frequency coverage, and jamming-to-signal ratio. Also, availability of spread spectrum modulation.

11.2 EXAMPLE #1: A FLOATING DATA-COLLECTION PLATFORM

The first application concerns the choice of a coding technique to meet specified performance requirements for a data-collection system consisting of 1000 to 10,000 floating (both fresh and salt water) data collection platforms which periodically transmit a burst of data from each platform through a geosynchronous satellite to a common receiving and processing station on earth. This transmission is at a rate of 100 b/s for 10 to 20 seconds. Thus, if error-control coding is to be used, the encoder must be inexpensive, so that 1000 to 10,000 such encoders will not cost the user a small fortune.

The data were to be transmitted using an antenna whose pattern is shown in Figure 11.1. The gain from this pattern is at most 3 to 4 dB relative to that of a perfectly spherical (isotropic) antenna pattern; i.e., 3 to 4 dBi. This type of coverage was specified by the user. While giving essentially complete hemispherical coverage, the pattern has two serious disadvantages, both of which lead to nearly the same result as far as received signal strength at the satellite is concerned. Note in Figure 11.1 that there is a rapid drop in gain both along the pattern's axis of symmetry and near the plane normal to this axis. Thus, any time either of these low-gain portions of the pattern is oriented toward the satellite during transmission as the result of motion by the floating platform, low signal strength or fading occurs. Even if platform motion does not occur, the phenomenon of multiple propagation paths (multipath) can produce fading at the satellite as a result of destructive interference between the direct and reflected signals.

At the time, the best available experimental results on fading channels for the band to be used showed 5 to 6 dB average and 10 dB maximum fades in one case, and maximum fades of 12 dB lasting anywhere from 5 to 20 seconds in another case. There was strong evidence to suggest that the broad, deep fades were caused primarily by multipath.

The challenge to error-control can now be seen: at 100 b/s, even a 5-second fade will affect 500 bits. Thus, a rate-1/2 code would have to be able to correct 1000-bit bursts. More fundamental in this situation is the fact that 5 seconds represents 25 to 50% of a single transmission. In fact, it was concluded that for fades of two seconds or more, a platform's transmission would have to be repeated or lost.

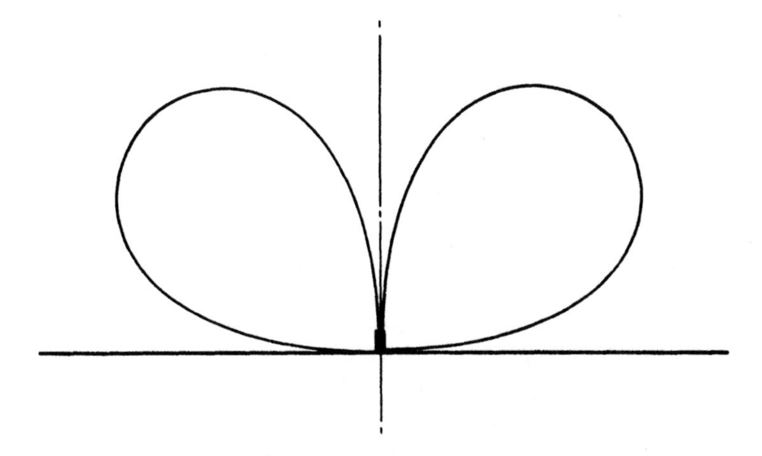

Figure 11.1 Antenna pattern for floating data-collection platform.

During intervals between fades, the channel between each floating platform and the satellite is the one assumed to exist at most times between any earth station and a satellite; i.e., the Gaussian channel. The average bit error probability P_B for this channel was estimated based on one of the sets of experiments from which the fading channel statistics just discussed were extrapolated. For tests in which there were bit errors, P_B ranged from 3×10^{-5} to 3×10^{-2}; these results included effects of any fading that may have occurred.

With this description of the physical aspects of the situation and of the data for what was believed to be a channel with behavior very similar to the one actually being studied, the following question was posed: what is the best solution for coping with bursts of errors caused by deep fades while providing protection against randomly occurring errors, all within the additional constraints of lowest possible cost, physically small size, and low power consumption? The following error-control (four correction and one detection) schemes were considered in this light:

- interleaved BCH codes;
- nonbinary codes;
- sequential decoding;
- diffuse threshold decoding;
- error detection with retransmission.

Interleaved BCH codes were considered by reviewing some simulations performed on data from a link characterized by impulse noise, frequency-selective fading, and in-band interference. Recall that interleaving spreads out the effect of a burst by rearranging code symbols. The average P_B was improved by about two orders of magnitude over the raw (unencoded) value of nearly 10^{-2} by the use of a (63, 36) five-error-correcting BCH code interleaved 1027 times. Unfortunately, the total number of data bits necessary in this case is 1027 × 36, or about 37,000, far in excess of the 1000 to 2000 data bits per block in the application.

The nonbinary codes considered used symbols consisting of m binary digits each. These codes had already been extensively tested via computer simulation and actual use. One code considered had $m = 8$, block length of 2040 binary digits, and the ability to correct up to 120 binary digit errors. At the time of the study, the encoder for this code was estimated to cost $1350 in a buy of 10 units, a figure well in excess of the allowable cost. Another code, having $m = 4$ with $k = 7$ and $n = 15$ 4-bit symbols, was also examined for its performance when interleaved 33 times, and for its cost. The overall interleaved block length of 924 data bits was compatible with the message length. Its burst-correcting capability at 100 b/s is about 2.4 seconds of channel digits, which is very good indeed. Decoder cost was estimated at $13,000.

Sequential decoding was judged ineffective against the long bursts anticipated because of the long searches and consequent input buffer overflows. (See Chapter 9 and Section 11.2 for additional discussion.)

Diffuse threshold decoding was found to provide protection against up to three seconds of errors with rate = 1/2 constraint length of 1200 to 1800 channel bits, and a guard space of about 1800 channel bits between bursts. Encoder parts cost was estimated at $100, and the decoder was available commercially for about $5000.

Error detection and retransmission were found to provide protection against fades up to about one second, as well as against random errors. One potential drawback was the need to add retransmission logic to one class of platform, thus increasing the cost and complexity.

On balance, the recommendation was to use either the interleaved $(n, k) = (15, 7)$, $m = 4$, code or diffuse threshold decoding.

Today this same set of requirements probably could be met by a commercially available Reed-Solomon encoder-decoder combination having $(n, k) = (31, 15)$ 5-bit symbols and the ability to correct up to eight symbol errors. For the 1000-bit transmission, interleaving to a depth of 14 would provide an adequate number of data bits and a burst-correcting capability of 560 digits or nearly 3 seconds at 100 b/s.

11.3 EXAMPLE #2: SATELLITE AND SPACE COMMUNICATION

Heller and Jacobs [11.1] treated not only code choice but also the choice of modulation in their analysis of satellite and space communication channels; these could be between satellites, between satellite and ground, or between space probe and ground. The channel model used happens not only to fit observed behavior very well but also to be the one most often assigned because of its analytical tractability, the additive white Gaussian noise (AWGN) channel. The whiteness of the noise implies independence of noise statistics from one bit interval to the next. At the same time, the channel response to the signal is sufficiently broadband to permit the modest bandwidth expansion required by the error-control technique.

Two algorithms are considered for this memoryless channel: Viterbi decoding and sequential decoding. You may recall that both of these algorithms were originally developed, as were many convolutional and block codes, for a memoryless channel. For a space-based system, a key constraint is the need to minimize the size and weight of the equipment, including the transmitter in particular.

Reduction in transmitter power means a corresponding reduction in received E_B/N_0, which, of course, leads directly to an increase in raw, uncoded bit error probability P_B. Possible courses of action involve some combination of the following:

- accept a higher P_B;
- recoup ERP by increasing transmitting antenna gain;
- increase received power and E_B by increasing receiving antenna gain;
- seek the optimum modulation of the carrier;
- use error control.

The last two alternative approaches are chosen for further analysis, with the condition that efficient, though not necessarily optimum, modulation be chosen. This turns out to be binary phase-shift keying (BPSK). Three main reasons are given for this choice:

1. BPSK waveforms are easy to generate. Furthermore, their amplification in a TWT amplifier makes best use of the TWT efficiency near saturation, compared to multilevel amplitude-modulated waveforms.
2. BPSK is nearly optimum in terms of E_B/N_0 *versus* P_B (9.6 dB at $P_B = 10^{-5}$). (Only orthogonal signaling is better.)
3. BPSK modulation of two quadrature carriers is equivalent to quadrature phase modulation of a single carrier.

Viterbi decoding was examined for both hard and soft (4-level and 8-level) received digit decisions. Emphasis was on codes of rate $R = 1/2$ and lengths $K = 3$ through 8, although rates 1/3 and 2/3 were also studied. In all cases, 3-level quantization gave an added improvement of about 2 dB at $P_B = 10^{-5}$, compared with hard decisions. The price of the 0.3

to 0.5 dB improvement in rate-1/3 codes compared with rate-1/2 codes is increased bandwidth and increased difficulty in symbol (channel digit) tracking due to decreased symbol energy to noise ratio E_s/N_0, where $E_s = RE_B$. At the time this study was performed, a Viterbi decoder having $R = 1/2$ and $K = 7$ and operating at up to 2 megabits per second with 2-, 4-, or 8-level quantization was cited as being available off the shelf in a model that required a total of 356 TTL integrated circuits. This relatively simple decoder provided more than 5 dB improvement in E_B/N_0 at $P_B = 10^{-5}$ compared to uncoded BPSK, and 6 dB at $P_B = 10^{-8}$, using 8-level decisions.

Heller and Jacobs [11.1] pointed out that the output P_B for either type of decoding is extremely low in the absence of bursts of errors at the decoder input. When such bursts do occur, the effect on a sequential decoder is far more serious than it is on a Viterbi decoder. In the former case, decoder output errors are likely to result not as a direct consequence of the decoder's inability to handle bursts, but rather because of the following chain of causally related events:

- burst of decoder input errors;
- long searches for correct path;
- decoder input buffer overflow;
- loss of received data;
- undetected errors at decoder output.

In the case of the Viterbi decoder, however, error bursts at the input will cause error bursts at the output of only one to three constraint lengths, with no loss of data, because the decoder produces decisions at a fixed rate independent of channel noise.

While soft decisions definitely benefit the Viterbi decoder performance with negligible impact on decoder size and processing time, this is not the case with sequential decoding, where there is a considerable price, both in terms of processing time and in branch metric storage. Consequently, the use of soft decisions with sequential decoding is not recommended. As Heller and Jacobs [11.1] noted, the performance of the Viterbi decoder degrades gracefully with adverse channel or receiver conditions.

Chapter 12
Synchronization Codes

12.1 BACKGROUND

Synchronization can be defined as the process of bringing two entities into time agreement within some tolerance. In communication there are at least five levels of synchronization beginning with the very small time increments (often nanoseconds or less) associated with clock synchronization and extending through some combination of chip, bit, byte, or branch, and frame, message, or packet synchronization. Each of these levels involves adjustments at a single receiver to agree with the corresponding level in the received signal. The levels of synchronization just listed constitute a hierarchy in the sense that each level in the list must be achieved before any of the subsequent levels listed can be achieved.

Network synchronization is a somewhat different process in that a number of terminals (transmitters and/or receivers) must agree in time, at least to the point of knowing the values of the time differences between any terminal and each of the others.

Clock "sync" must be achieved to within some fraction of a cycle, (in many cases, the clock function is embodied in a carrier waveform), or else chip sync (or bit sync, if there are no chips) cannot be accomplished. Similar statements apply to each pair of successively higher orders of synchronization. Even when the boundaries of message or packet, branch or byte, bit, chip, and carrier cycle have all been matched up, there remains the problem of time ambiguity, i.e., alignment could be off by an amount equal to an integer multiple of the message length (or whatever the largest unit happens to be).

There are two main criteria for determining whether synchronization has been achieved. These criteria are sometimes used singly and sometimes together. The most common criterion is the autocorrelation function $R(\tau)$ of the synchronizing sequence, where τ is the offset between two copies or versions of the sequence (usually one received and one locally generated copy). In the ideal case, $R(\tau)$ has a value at zero offset that is very large compared to its value at any other offset, and this latter value of $R(\tau)$ is zero. This situation is actually achieved for some sequences, but more often a value equal in magnitude to a very small fraction of the correlation peak is the case. The second criterion is Hamming distance. Most of the discussion of Hamming distance in Chapter 3 applies directly to the synchronization problem.

The length of a synchronization sequence can range from as little as 2 or 3 in the case of the shortest Barker codes to a few hundred. The choice of length will depend largely on the application. Bear in mind that interference, in the form of noise, jamming, or other users, is often present. Consequently, in all cases the guiding principle is to use as short a sync sequence as possible without having an unacceptably high probability of either false acquisition (i.e., synchronization at the wrong point of the desired message or, worse yet, synchronization with the wrong message) or missed acquisition. The need for a short sync time may simply be driven by the desire for maximum efficiency, or it may be as serious a matter as minimizing the period of vulnerability prior to synchronization with a subsequent sequence which is used to spread the signal spectrum for antijamming (AJ) purposes.

12.2 SCOPE OF THE CHAPTER

Techniques of various sorts exist for achieving synchronization at the different levels discussed in the preceding section. Of primary interest here are the techniques based on codes that are used to synchronize at the chip, bit, and message levels. The well-known maximum-length or other pseudonoise (PN) codes, produced by linear-feedback shift registers with certain specific tap connnections, are used at all three of these levels. In certain situations it is just as effective and more efficient to use a Barker code. Both PN and Barker codes, when used for synchronization, are usually inserted into the data stream with a resultant (and usually small) decrease in efficiency. (An alternative procedure is to transmit the sync code over a separate channel, e.g., at another carrier frequency.) A conceptually different approach is to use encodings of the data words or symbols that have self-synchronizing properties. This involves the use of encoding schemes in which each word is preceded by a fixed pattern (called a *comma*) or in which no code word is allowed to be the same as the beginning of another code word (called a *prefix code* or *comma-free code*).

It is important to bear in mind that nearly all codes used for synchronization have no error-control capability. There are a few self-synchronizing error-correcting codes, and there are codes that can correct the insertion or deletion of one or a few bits. These will not be discussed here.

12.3 PN CODES

A PN code is generated by a shift register (SR) in which the modulo-2 sum of the contents (0 or 1) of certain SR stages is fed back as the newest bit in the SR. An example, a 5-stage SR, is shown in Figure 12.1. The shift register must be initialized so that at least one stage contains a 1. (If this were not done, then clearly the feedback would always be 0 and no useful output would result.) Successive contents of the last stage are the SR output. Note that the use of a shift register to generate a PN sequence differs from its use as an encoder for a cyclic block code or a convolutional code in that the error-control code application involves at least one input sequence during SR operation whereas the only input for a PN generator is the initial SR contents. The two applications do have a common analytical basis in the use of properties of polynomials over a Galois field. In particular, for an SR of m stages, the maximum period of the output is $2^m - 1$, achieved only when the feedback connections correspond to a primitive polynomial of degree m. (See Chapter 4 for a discussion of primitive polynomials and relevant mathematical background.) The SR of Figure 12.1 corresponds to the primitive polynomial $1 + x^2 + x^5$. (Tables of primitive polynomials are in the books by Lin and Costello [5.8] and Holmes [12.1], among others.) This section considers SRs with outputs that are maximum-length or *maximal* sequences.

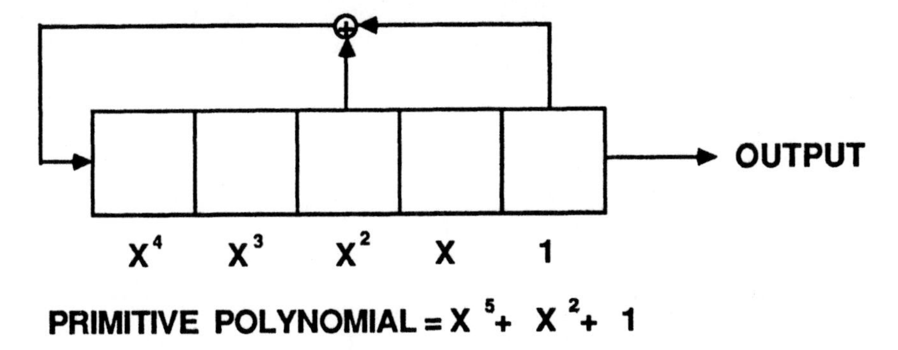

Figure 12.1 A maximal shift register generator where $m = 5$.

Maximal sequences have some important properties. These can be derived easily, but are stated here without proof. The derivations can be found in any of several excellent texts, including those by Torrieri [12.2], Holmes [12.1], and Peterson and Weldon [3.2]. The following properties are of greatest use in the study of maximal PN sequences for synchronization:

1. Each nonzero initial state of the SR gives an output sequence that is a cyclic shift of any other output sequence.
2. The sum of any two shifts of the output sequence is yet another shift of the same sequence.
3. The number of 1s is always 1 greater than the number of 0s in a maximal sequence (i.e., 2^{n-1} 1s, $2^{n-1} - 1$ 0s).
4. The autocorrelation function over a full period has the value of 1 for zero offset and $-1/(2^m - 1)$ for all other offsets. (The autocorrelation function for a PN sequence is computed from $R(\tau) = (A - D)/(A + D)$, where A = number of agreements between the two copies of the PN sequence, D = number of disagreements, $A + D = 2^m - 1$, $R(\tau)$ is defined at the discrete values 0, $\pm T_c$, $\pm 2T_c, \ldots$, and T_c is the duration of a digit (usually called a *chip*) in the PN sequence.

Property 4, which actually follows almost trivially from property 2, is the basis for the use of PN sequences in synchronization. That is, the sharp peak at $\tau = 0$ and the near-zero value at $\tau \neq 0$ even at modest values of m make it easy to determine when synchronization has been achieved. Table 12.1 shows one complete period of output from the SR generator of Figure 12.1, starting with the initial contents of Figure 12.1. Figure 12.2 shows the autocorrelation function of $R(\tau)$ for the sequence of Table 12.1. The period is $2^5 - 1 = 31$, so at offsets (including 0) that are not integer multiples of 31, $R(\tau) = -1/31$. Note that it is customary to connect the discrete points with an unbroken line.

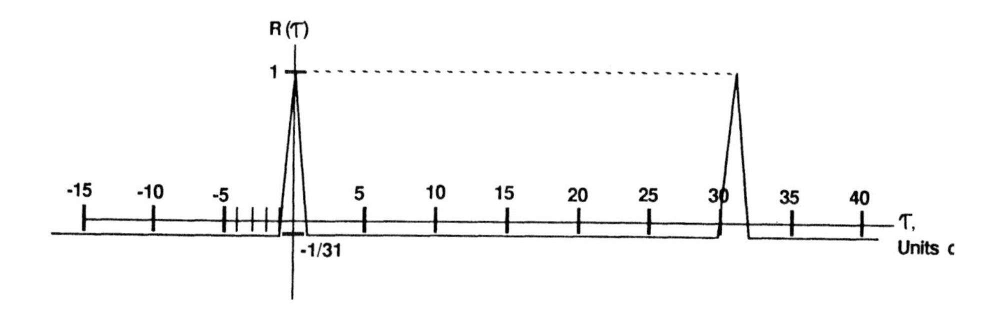

Figure 12.2 Autocorrelation function for the PN sequence of Table 12.1.

Table 12.1 Output of shift register generator of Figure 12.1
$p(x) = x^5 + x^2 + 1$

Time	*Stage*				
	X^4	X^3	X^2	X	1
0	0	0	1	0	0
1	1	0	0	1	0
2	0	1	0	0	1
3	1	0	1	0	0
4	1	1	0	1	0
5	0	1	1	0	1
6	0	0	1	1	0
7	1	0	0	1	1
8	1	1	0	0	1
9	1	1	1	0	0
10	1	1	1	1	0
11	1	1	1	1	1
12	0	1	1	1	1
13	0	0	1	1	1
14	0	0	0	1	1
15	1	0	0	0	1
16	1	1	0	0	0
17	0	1	1	0	0
18	1	0	1	1	0
19	1	1	0	1	1
20	1	1	1	0	1
21	0	1	1	1	0
22	1	0	1	1	1
23	0	1	0	1	1
24	1	0	1	0	1
25	0	1	0	1	0
26	0	0	1	0	1
27	0	0	0	1	0
28	0	0	0	0	1
29	1	0	0	0	0
30	0	1	0	0	0
31	0	0	1	0	0

Maximal PN sequences are cheap and easy to generate, and the generators can now be made programmable for a wide range of sequence lengths and choice of sequences of the same length. The only potential disadvantage is the cross-correlation between distinct sequences of the same length, which can be a problem in a multiuser environment. One possible alternative in this situation is the use of Barker codes (discussed in the next section). If the sync sequence is to be sent several times in succession, then a maximal sequence is optimum. If the sync sequence is to be sent only intermittently, then a Barker code is a superior alternative.

12.4 BARKER CODES

Barker codes were first described by R.H. Barker in 1953 [12.3]. They are a very small set of binary sequences with outstanding full-length and partial-length autocorrelation properties. The entire set of known Barker codes is shown in Table 12.2. It has been proven that there are no Barker codes of odd length greater than 13, and that if any others of even length exist, the length must be the square of an integer. Since no Barker codes of even length other than 2 and 4 have been found, there is strong reason to believe that there are no undiscovered Barker codes. The Barker codes of lengths 3 and 7 are also maximal sequences.

Because a Barker code is usually employed without repetition of the code sequence, the autocorrelation function for nonzero shifts is computed for successively smaller overlaps of length $N - (\tau/T_c)$, where N is the code length and T_c is the chip duration. Because of this manner of computing $R(\tau)$, it is not meaningful to normalize by dividing by the sequence length. The result is, therefore, simply,

$$R(\tau) = A - D, \quad \tau = 0, \pm T_c, \ldots, \pm NT_c,$$

where A = number of agreements and D = number of disagreements, just as for maximal sequences. Consequently, $R(\tau)$ for Barker codes has a peak value of N at $\tau = 0$, and alternates between 0 and -1 for odd values of N. For even N (i.e., $N = 2$ or 4), the behavior of $R(\tau)$ is slightly different from this. (Try it and see!) Figure 12.3 shows part of the computation and graph of $R(\tau)$ for the Barker code of length 7.

Table 12.2 Barker Codes

Length	*Code*
2	00
3	001
4	0001 or 0010
5	00010
7	0001101
11	00011101101
13	0000011001010

Sequence = 0001101 Length = 7

A: Agreement D: Disagreement

$$R(\tau) = n(A) - n(D)$$

Offset $\tau = 1$:

```
0001101   n(A) = 3 = n(D)
 000110
```

Correlation Window $R(\tau) = 0$

$\tau = 2$:

```
0001101   n(A) = 2, n(D) = 3
  00011   R(τ) = −1
```

$\tau = 3$:

```
0001101   n(A) = 2 = n(D)
   0001   R(τ) = 0
```

$\tau \geq 7$: $n(A) = n(D) = R(\tau) = 0$

(a) Computation of R (τ)

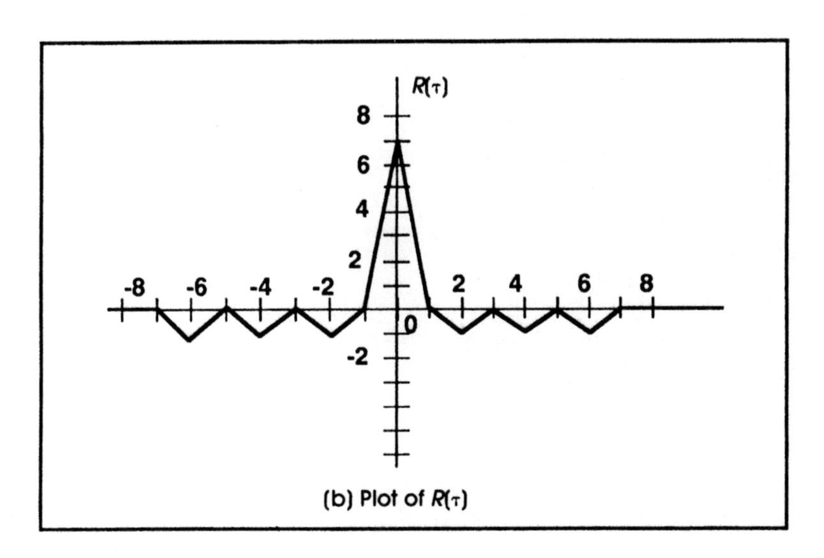

(b) Plot of $R(\tau)$

Figure 12.3 Barker code autocorrelation.

Appendix
Introduction to Matrices

A.1 DEFINITIONS AND NOTATION

A *matrix* is a rectangular array of entries or elements; these may be elements of a field, a ring, or any of several other algebraic systems. Because our main interest here is in elements of a field, usually finite, descriptions and examples will pertain only to matrices of field elements. The operations such as addition and multiplication defined for simpler algebraic structures are also defined for matrices; furthermore, because of their greater complexity, matrices have other operations defined.

Row and *column* are used with their everyday meanings to indicate a particular subset of the elements of a matrix. A matrix is usually denoted by a capital Roman letter, with any of its elements denoted by a subscripted corresponding lower-case letter. The subscripts are ordered; that is, the first indicates the row and the second, the column in which that element is located in the matrix. Thus $a_{2,4}$ is the element from the second row and fourth column of matrix A, with A often written as:

$$A = \begin{bmatrix} a_{1,1} & a_{1,2} & a_{1,3} & \cdots & a_{1,n} \\ a_{2,1} & a_{2,2} & a_{2,3} & \cdots & a_{2,n} \\ \cdot & \cdot & \cdot & & \cdot \\ \cdot & \cdot & \cdot & & \cdot \\ \cdot & \cdot & \cdot & & \cdot \\ a_{m,1} & a_{m,2} & a_{m,3} & \cdots & a_{m,n} \end{bmatrix} \tag{A.1}$$

Another way of denoting a matrix is by enclosing a general element in parentheses. Thus,

$$A = (a_{i,i})$$

A matrix having m rows and n columns is called an $m \times n$ matrix. If $m = n$, the matrix is *square*. The elements $a_{i,j}$ in any square matrix constitute the *main diagonal* of the matrix. The *transpose* A^T of a matrix A is defined by $(a_{i,j})^T = (a_{j,i})$. That is, A^T is formed from A by letting the ith row of A become the ith column of A^T. Thus, if

$$A = \begin{bmatrix} 1 & 2 & 3 \\ 4 & 5 & 6 \end{bmatrix},$$

then

$$A^T = \begin{bmatrix} 1 & 4 \\ 2 & 5 \\ 3 & 6 \end{bmatrix}$$

In particular, the transpose of a row vector is a column vector, and *vice versa*. This property is used extensively in the formulation of linear codes and their properties.

Finally, if

$$A = \begin{bmatrix} \alpha_1 \\ \alpha_2 \end{bmatrix}$$

with

$$\alpha_1 = (1, 2, 3)\ ,\ \alpha_2 = (4, 5, 6)$$

then

$$\alpha_1^T = \begin{bmatrix} 1 \\ 2 \\ 3 \end{bmatrix}, \quad \alpha_2^T = \begin{bmatrix} 4 \\ 5 \\ 6 \end{bmatrix}$$

and A^T may be written

$$A^T = [\alpha_1^T\ \alpha_2^T]$$

A.2 ELEMENTARY OPERATIONS ON MATRICES

Addition of two or more matrices can be carried out if, and only if, the matrices have the same number of rows and the same number of columns. In this case, matrix addition is defined in terms of the addition of corresponding elements, using the operations of the field of which the elements are members. That is, if matrices A and B are both $m \times n$ with elements from the same field F, then the sum of A and B exists and is given by

$$\mathrm{A} + \mathrm{B} = (\mathrm{a}_{i,j}) + (\mathrm{b}_{i,j}) \tag{A.2}$$

with each sum $\mathrm{a}_{i,j} + \mathrm{b}_{i,j}$ formed in F.

From earlier work with fields, you might guess (correctly) that if subtraction is defined for the field elements, then it is also defined for the corresponding matrices. Note in passing that the additive identity matrix is the one whose elements are all zeroes; it will be denoted by Φ.

Multiplication of any matrix by a field element (scalar) is defined exactly as it was in the case of a vector:

$$\mathrm{cA} = (\mathrm{ca}_{i,j}) \tag{A.3}$$

Another kind of multiplication is also defined for matrices; this is the product of two matrices. It is necessary that the number of columns in the first (left-hand) factor be equal to the number of rows in the second (right-hand) factor. That is, if the first matrix is $m \times n$, then the second must be $n \times p$; the product is then an $m \times p$ matrix. (Note: any two, or all three, of m, n, and p can be equal.)

If $\mathrm{c}_{i,j}$ is the general element of the product C = AB, then $\mathrm{c}_{i,j}$ is defined by

$$\mathrm{c}_{i,j} = \sum_{\ell=1}^{n} \mathrm{a}_{i,\ell}\mathrm{b}_{\ell,j} \qquad i = 1, \ldots, m \qquad j = 1, \ldots, p \tag{A.4}$$

Example A.1. Let

$$\mathrm{A} = \begin{bmatrix} 4 & 2 \\ -1 & 1 \end{bmatrix}, \quad \mathrm{B} = \begin{bmatrix} 1 & 0 & -9 \\ 7 & -2 & 5 \end{bmatrix}, \quad \mathrm{C} = \begin{bmatrix} -3 & 5 \\ 1 & -2 \end{bmatrix}$$

where the integer elements of matrices A, B, and C are regarded as coming from the (infinite) field of rational numbers. Then A + B and B + C are not defined, but

$$\mathrm{A} + \mathrm{C} = \begin{bmatrix} 4 + (-3) & 2 + 5 \\ -1 + 1 & 1 + (-2) \end{bmatrix} = \begin{bmatrix} 1 & 7 \\ 0 & -1 \end{bmatrix}$$

and

$$2B = 2\begin{bmatrix} 1 & 0 & -9 \\ 7 & -2 & 5 \end{bmatrix} = \begin{bmatrix} (2)(1) & (2)(0) & (2)(-9) \\ (2)(7) & (2)(-2) & (2)(5) \end{bmatrix}$$

$$= \begin{bmatrix} 2 & 0 & -18 \\ 14 & -4 & 10 \end{bmatrix}$$

The product AB is defined because the number of columns in A equals the number of rows in B:

$$P = AB = \begin{bmatrix} 4 & 2 \\ -1 & 1 \end{bmatrix} \begin{bmatrix} 1 & 0 & -9 \\ 7 & -2 & 5 \end{bmatrix}$$

$$= \begin{bmatrix} (4)(1) + (2)(7) & (4)(0) + (2)(-2) & (4)(-9) + (2)(5) \\ (-1)(1) + (1)(7) & (-1)(0) + (1)(-2) & (-1)(-9) + (1)(5) \end{bmatrix}$$

$$= \begin{bmatrix} 18 & -4 & -26 \\ 6 & -2 & 14 \end{bmatrix} \qquad (A.5)$$

Next, compute

$$AC = \begin{bmatrix} 4 & 2 \\ -1 & 1 \end{bmatrix} \begin{bmatrix} -3 & 5 \\ 1 & -2 \end{bmatrix} = \begin{bmatrix} -10 & 16 \\ 4 & -7 \end{bmatrix} \qquad (A.6)$$

and

$$CA = \begin{bmatrix} -3 & 5 \\ 1 & -2 \end{bmatrix} \begin{bmatrix} 4 & 2 \\ -1 & 1 \end{bmatrix} = \begin{bmatrix} -17 & -1 \\ 6 & 0 \end{bmatrix} \qquad (A.7)$$

Obviously, $AC \neq CA$, illustrating the property that matrix multiplication is not in general commutative.

There is a multiplicative identity I. It is a square matrix having all 1s on its main diagonal and 0s elsewhere. The usual notation is I_k for the $k \times k$ identity matrix. If the rows of a square matrix A are linearly independent, then A has a multiplicative inverse A^{-1} which satisfies

$$A\,A^{-1} = A^{-1}A = I$$

For example, suppose that $A = \begin{bmatrix} 2 & 1 \\ 1 & 4 \end{bmatrix}$ and let

$$A^{-1} = B = \begin{bmatrix} b_{1,1} & b_{1,2} \\ b_{2,1} & b_{2,2} \end{bmatrix}$$

Then AB = I implies that

$$2b_{1,1} + b_{2,1} = 1 \qquad b_{1,1} + 4b_{2,1} = 0 \tag{A.8}$$

$$2b_{1,2} + b_{2,2} = 0 \qquad b_{1,2} + 4b_{2,2} = 1$$

This system of equations can be solved to give

$$b_{2,1} = -1/7 \quad b_{1,1} = 4/7 \quad b_{1,2} = -1/7, \quad b_{2,2} = 2/7$$

You should verify that these values satisfy AB = BA = I.

Not all square matrices have a multiplicative inverse. For example, if $A = \begin{bmatrix} 1 & 0 \\ 2 & 0 \end{bmatrix}$, then AB = I leads to the equations:

$$\begin{aligned} b_{1,1} &= 1 \qquad b_{1,2} = 0 \\ 2b_{1,1} &= 0 \qquad 2b_{1,2} = 1 \end{aligned} \tag{A.9}$$

a system that obviously has no solution. Thus, there is no matrix B that satisfies AB = I in this case, and A is called a *singular* matrix.

A.3 CALCULATION OF THE INVERSE

The technique applied at the end of Section A.2 is not the best way to calculate the inverse of a matrix (also called *inverting a matrix*) especially for matrices larger than 2 by 2. The preferred method, and one that is also used with some digital computer matrix inversion programs, consists of transforming the given matrix into the identity matrix and at the same time applying to an identity matrix the same operations that are applied to the given matrix. When this series of operations is complete, the identity will have been transformed to the inverse of the original matrix. A compact way of doing this is to define M = [A|I] and operate on rows of M until you get $M' = [I|A^{-1}]$. Before illustrating this technique, it will be helpful to define an *elementary row transformation* of a matrix as any of the following three operations on the rows of a matrix:

(R1) The interchange of two rows;

(R2) Multiplication of a row by a nonzero scalar (here, as with vectors, the term *scalar* refers to a field element);

(R3) Addition to a row of a nonzero scalar multiple of another row.

Example A.2. Let A be defined over the real numbers as follows:

$$A = \begin{bmatrix} 1 & 2 & 1 \\ 2 & 1 & 2 \\ 0 & 1 & 1 \end{bmatrix}$$

Then operation (R1) applied to rows 1 and 3 gives

$$A_1 = \begin{bmatrix} 0 & 1 & 1 \\ 2 & 1 & 2 \\ 1 & 2 & 1 \end{bmatrix}$$

Operation (R2) applied to row 1 of A, for the scalar equal to 4, gives

$$A_2 = \begin{bmatrix} 4 & 8 & 4 \\ 2 & 1 & 2 \\ 0 & 1 & 1 \end{bmatrix}$$

Finally, operating on A with (R3) by adding (-1) times row 3 to row 1 gives

$$A_3 = \begin{bmatrix} 1 & 1 & 0 \\ 2 & 1 & 2 \\ 0 & 1 & 1 \end{bmatrix}$$

An important observation about these row transformations is this: each row transformation of matrix A can be obtained by left-multiplying A by the matrix obtained by applying the corresponding row transformation to I_3. (Left- (or pre-) multiplying matrix A by matrix P means forming the product PA.) Thus,

$$A_1 = U_1 A \tag{A.10}$$

where

$$U_1 = \begin{bmatrix} 0 & 0 & 1 \\ 0 & 1 & 0 \\ 1 & 0 & 0 \end{bmatrix}$$

was obtained by applying transformation (R1) to rows 1 and 3 of I_3. (You may want to verify (A.10) by carrying out the indicated matrix multiplication.)

Similarly,

$$A_2 = V_1A \qquad \text{for } V_1 = \begin{bmatrix} 4 & 0 & 0 \\ 0 & 1 & 0 \\ 0 & 0 & 1 \end{bmatrix}$$

(where row 1 of I_3 was multiplied by 4 to get V_1);
and

$$A_3 = W_1\,A \qquad \text{for } W_1 = \begin{bmatrix} 1 & 0 & -1 \\ 0 & 1 & 0 \\ 0 & 0 & 1 \end{bmatrix}$$

(row 3 of I_3 subtracted from row 1 to get W_1).

Let P be the product of all transformation matrices which, when applied in succession to a $k \times k$ matrix A, produce I_k:

$$PA = I_k \tag{A.11}$$

Then, for M as defined earlier in this discussion:

$$PM = [PA|PI_k] = [I_k|P] \tag{A.12}$$

Equation (A.12) follows from (A.11) and from the fact that multiplication by the identity matrix leaves any matrix unchanged. Post-multiplying both members of (A.11) by A^{-1} gives

$$(PA)A^{-1} = I_kA^{-1} \tag{A.13}$$

By the associative law of multiplication, the left-hand side of (A.13) is just P, so that this equation reduces to

$$P = A^{-1}$$

which is the result promised earlier. You have now seen how to find the inverse (when it exists) of a matrix by the use of row transformations, which is an easy technique to understand and to carry out. A simple example should be useful at this point.

Example A.3.

$$\text{Let } A = \begin{bmatrix} 4 & 1 \\ 2 & 1 \end{bmatrix}. \text{ Then } M = \begin{bmatrix} 4 & 1 & 1 & 0 \\ 2 & 1 & 0 & 1 \end{bmatrix}$$

Assume that all computation is in the field of rational numbers. The following row operations are applied in succession to M and the matrices derived from it:

Add $(-1) \times$ row 2 to row 1.
Multiply row 1 (of matrix obtained in preceding step) by ½.
Add $(-2) \times$ row 1 to row 2.

The matrices that result are

$$\begin{bmatrix} 2 & 0 & 1 & -1 \\ 2 & 1 & 0 & 1 \end{bmatrix}, \begin{bmatrix} 1 & 0 & 1/2 & -1/2 \\ 2 & 1 & 0 & 1 \end{bmatrix},$$

$$\begin{bmatrix} 1 & 0 & 1/2 & -1/2 \\ 0 & 1 & -1 & 2 \end{bmatrix} = [I_2 \; A^{-1}]$$

Thus,

$$A^{-1} = \begin{bmatrix} 1/2 & -1/2 \\ -1 & 2 \end{bmatrix}$$

The relationships $AA^{-1} = A^{-1}A = I$ can be checked by the reader.

A.4 REDUCTION TO TRIANGULAR FORM

The sequence of steps used in transforming the matrix of Example A.3 was not chosen at random, although for very small matrices like A in this example, the sequence is not so important. There is a definite procedure to be followed in reducing a matrix to the identity matrix, or, more generally, to triangular form.

First, recall the following definition: the set of elements $a_{i,i}$ of the $m \times n$ matrix A is called the *main diagonal* of A, whether A is square or not. With this definition in mind, the upper triangular form of a matrix (so-called for reasons which should be clear from its properties) is defined as:

1. Every element on the main diagonal is 0 or 1.
2. Every element below the main diagonal is 0.

3. If an element on the main diagonal is 1, all other elements in that column are 0.

To put a matrix in this form, the following procedure, which uses only row transformations, is applied to the ith row, for $i = 1, 2, \ldots, m$ in succession:

Step 1. If $a_{i,i} = 0$, interchange the ith and kth rows (i.e., apply row transformation (R1) defined in Section A.3) for any $k > i$, so that the new $a_{i,i} \neq 0$. If this is impossible, repeat Step 1 for row $i + 1$. (If $i = m$, then omit Step 1.)

Step 2. If $0 \neq a_{i,i} \neq 1$, multiply all elements of row i by $a_{i,i}^{-1}$ (i.e., apply transformation (R2)).

Step 3. For $\ell \neq i$, add $-a_{\ell,i}$ times the ith row to the ℓth row (i.e., apply transformation (R3)).

Step 1 places a nonzero element on the main diagonal if that is possible for the row in question; otherwise, a 0 is placed. Step 2 scales a nonzero $a_{i,i}$ to 1. Step 3 places zeros in all positions above and below a nonzero $a_{i,i}$. This procedure is best understood with the help of an example.

Example A.4. Consider the matrix:

$$B = \begin{bmatrix} 0 & 1 & 1 & 2 \\ 1 & 4 & 3 & 6 \\ 0 & 3 & 1 & 0 \end{bmatrix}$$

Set $i = 1$ and observe that $b_{1,1} = 0$, while $b_{2,1} \neq 0$. Therefore, (Step 1) exchange rows 1 and 2 of B, giving

$$B_1 = \begin{bmatrix} 1 & 4 & 3 & 6 \\ 0 & 1 & 1 & 2 \\ 0 & 3 & 1 & 0 \end{bmatrix}$$

For B_1, Step 2 is unnecessary because the new $b_{1,1}$ is already equal to 1, while Step 3 is unnecessary because $b_{2,1} = b_{3,1} = 0$. Now set $i = 2$ and look at the second row of B_1. Because $b_{2,2} = 1$, Steps 1 and 2 are unnecessary. Step 3, however, must be carried out for $\ell = 1$ and $\ell = 3$. Thus, $b_{1,2} = 4$, so add $(-4) \times$ row 2 to row 1, giving

$$B_2 = \begin{bmatrix} 1 & 0 & -1 & -2 \\ 0 & 1 & 1 & 2 \\ 0 & 3 & 1 & 0 \end{bmatrix}$$

Similarly, $b_{3,2}$ is changed to 0 by adding $(-3) \times$ row 2 of B_2 to row 3:

$$B_3 = \begin{bmatrix} 1 & 0 & -1 & -2 \\ 0 & 1 & 1 & 2 \\ 0 & 0 & -2 & -6 \end{bmatrix}$$

In B_3, $b_{3,3} \neq 0$ and $i = m = 3$, so Step 1 is unnecessary. Because $b_{3,3} \neq 1$, Step 2, with $b_{3,3}^{-1} = -½$, gives

$$B_4 = \begin{bmatrix} 1 & 0 & -1 & -2 \\ 0 & 1 & 1 & 2 \\ 0 & 0 & 1 & 3 \end{bmatrix}$$

Finally, Step 3 applied to B_4 for $i = 3$, $\ell = 1$ and 2, gives

$$B_5 = \begin{bmatrix} 1 & 0 & 0 & 1 \\ 0 & 1 & 0 & -1 \\ 0 & 0 & 1 & 3 \end{bmatrix}$$

which is in upper triangular form.

A nonsquare matrix was intentionally chosen for this example to show how the idea of the main diagonal applies in such a situation. Notice that once all elements but one in a column have been made equal to 0, further row operations made according to the triangularization procedure will have no effect on elements in columns numbered less than the current value of i.

In the procedure just illustrated, any row that is a linear combination of others will ultimately become all 0s. Of what use is this? The number of nonzero rows remaining after triangularization is the number of linearly independent rows and hence the dimension of the vector space spanned by the rows of the matrix. The next example, based on the code used in Example 3.12, illustrates the statements made in this paragraph and the preceding one.

Example A.5. The details of this example are shown in Figure A.1. In this case, each row of the original matrix has been identified by means of a number enclosed in a circle. Operations on the rows are indicated throughout by using the numbers *originally* assigned to the rows. Thus, ⑦ − 2 × ④ means that twice the fourth row of the original matrix is subtracted from the seventh row. In the third stage of Figure A.1, parentheses are used to help identify operations on the matrix of the preceding stage. Each stage represents more than one of the steps of the triangularization process.

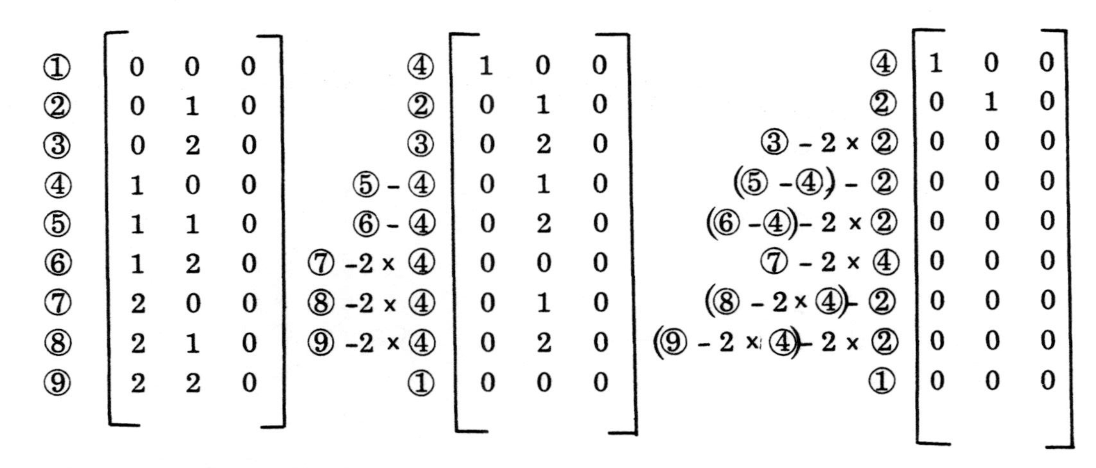

Note: All circled numbers refer to rows of first matrix. All computations are in GF(3), for which addition and multiplication tables appear in Section 3.5.

Figure A.1 Matrix reduction for code of Example 3.12.

Note that the final stage shows only two nonzero rows, clearly indicating that the entire code can be generated from the vectors (1, 0, 0) and (0, 1, 0).

Bibliography

[1.1] Glenn, A.B., "Multi-Service Communications Network for Civil Command and Control," *EASCON 76 Record,* September 26–29, 1976, pp. 24-A–24-G.

[1.2] Tanenbaum A.S., *Computer Networks,* Prentice-Hall, Englewood Cliffs, N.J., 1981.

[2.1] Wozencraft, J.M., and I.M. Jacobs, *Principles of Communication System Engineering,* John Wiley and Sons, New York, 1965.

[2.2] Carlson, A.B., *Communication Systems,* McGraw-Hill, New York, 1968.

[2.3] Haykin, S.S., *Communication Systems,* John Wiley and Sons, New York, 1983

[2.4] Korn, I., *Digital Communications,* Van Nostrand Reinhold, New York, 1985

[2.5] Cahn, C.R., "Comparison of Phase Tracking Schemes for PSK," *Proc. International Telemetering Conference,* 1971, pp. 172–180.

[2.6] Heller, J.A., "Sequential Decoding for Channels with Time Varying Phase," Ph.D. dissertation, Massachusetts Institute of Technology, Cambridge, MA, 1967.

[2.7] Cahn, C.R., G.K. Huth, and C.R. Moore, "Simulation of Sequential Decoding with Phase Locked Demodulation," *Magnavox Co. Report No. MX-TM-3119-71,* 1971.

[2.8] Viterbi, A.J., *Principles of Coherent Communication,* McGraw-Hill, New York, 1966.

[2.9] Gardner, F.M., *Phaselock Techniques,* John Wiley and Sons, New York, 1966.

[3.1] Gallager, R.G., *Information Theory and Reliable Communication,* John Wiley and Sons, New York, 1968.

[3.2] Peterson, W.W., and E.J. Weldon, Jr., *Error-Correcting Codes,* Second Ed., John Wiley and Sons, New York, 1972.

[3.3] MacWilliams, F.J., and N.J.A. Sloane, *The Theory of Error-Correcting Codes,* North-Holland, Amsterdam, 1977.

[3.4] Michelson, A.M. and A.H. Levesque, *Error-Contral Techniques for Digital Communications,* John Wiley and Sons, New York, 1985.

[5.1] Bose, R.C., and D.K. Ray-Chaudhuri, "On a Class of Error Correcting Binary Group Codes," *Information and Control,* Vol. 3, 1960, pp. 68–79.

[5.2] Bose, R.C., and D.K. Ray-Chaudhuri, "Further Results on Error Correcting Binary Group Codes," *Information and Control,* Vol. 3, 1960, pp. 279–290.

[5.3] Hocquenghem, A., "Codes correcteurs d'erreurs," *Chiffres,* Vol. 2, 1959, pp. 147–156.

[5.4] Berlekamp, E.R., *Algebraic Coding Theory,* McGraw-Hill, New York, 1968.

[5.5] Massey, J.L., "Shift Register Synthesis and BCH Decoding," *IEEE Trans. Information Theory,* Vol. IT-15, No. 1, January 1969, pp. 122–127.

[5.6] Carhoun, D.O., T.F. Roome, and E.A. Palo, "Finite Field Arithmetic with Charge Transfer Devices," *Third Int. Conf. Applications of Charge-Coupled Devices,* Edinburgh, Scotland, September 28–30, 1976.

[5.7] Carhoun, D.O., T.F. Roome, and E.A. Palo, "Error Correction Coding with Charge Transfer Devices," *Third Int. Conf. Applications of Charge-Coupled Devices,* Edinburgh, Scotland, September 28–30, 1976.

[5.8] Lin, S., and D. Costello, *Error-Control Coding: Theory and Applications,* Prentice-Hall, Englewood Cliffs, NJ, 1981.

[5.9] Stenbit, J.P., "Tables of Generators for Bose-Chaudhuri Codes," *IEEE Trans. Information Theory,* Vol. IT-10, No. 4, October 1964, pp. 390–391.

[5.10] Lin, S., *An Introduction to Error-Correcting Codes,* Prentice-Hall, Englewood Cliffs, NJ, 1970.

[5.11] Chien, R.T., "Block-Coding Techniques for Reliable Data Transmission," *IEEE Trans. Communication Technology,* Vol. COM-10, No. 5, October 1971, pp. 743–751.

[5.12] McEliece, R.J., *The Theory of Information and Coding,* Addison-Wesley, Reading, MA, 1977.

[5.13] Stenbit, J.P., *op. cit.* [5.9], p. 391

[5.14] Reed, I.S., and G. Solomon, "Polynomial Codes over Certain Finite Fields," *J. Society for Industrial and Applied Mathematics,* Vol. 8, 1960, pp. 300–304.

[5.15] Gorenstein, D., and N. Zierler, "A Class of Error-Correcting Codes in p^m Symbols," *J. Society for Industrial and Applied Mathematics,* Vol. 9, 1961, pp. 207–214.

[6.1] Wozencraft, J.M., and B. Reiffen, *Sequential Decoding,* The MIT Press, Cambridge, MA, 1961.

[6.2] Massey, J.L., "Shift Register Synthesis and BCH Decoding," *IEEE Trans. Information Theory,* Vol. IT-15, No. 1, January 1969, pp. 122–127.

[6.3] Heller, J.A., "Sequential Decoding: Short Constraint Length Convolutional Codes," Jet Propulsion Laboratory, California Institute of Technology, Pasadena, *Space Program Summary 37-54,* Vol. 3, 1968, pp. 171–174.

[6.4] Fano, R.M., *Transmission of Information,* published jointly by The MIT Press, Cambridge, MA, and John Wiley and Sons, New York, 1961.

[6.5] Larsen, K.J., "Short Convolutional Codes with Maximal Free Distance for Rates 1/2, 1/3, and 1/4," *IEEE Trans. Information Theory,* Vol. IT-19, No. 3, 1973, pp. 371–372.

[6.6] Odenwalder, J.P., "Optimal Decoding of Convolutional Codes," Ph.D. dissertation, Department of System Sciences, School of Engineering and Applied Science, University of California at Los Angeles, 1970.

[6.7] Bahl, L.R., and F. Jelinek, "Rate 1/2 Convolutional Codes with Complementary Generators," *IEEE Trans. Information Theory,* Vol. IT-17, No. 6, 1971, pp. 718–727.

[6.8] Paaske, E., "Short Binary Convolutional Codes with Maximal Free Distance for Rates 2/3 and 3/4," *IEEE Trans. Information Theory,* Vol. IT-20, No. 5, 1974, pp. 683–689.

[6.9] Johannesson, R., "Robustly Optimal Rate One-Half Binary Convolutional Codes," *IEEE Trans. Information Theory,* Vol. IT-21, No. 4, 1975, pp. 464–468.

[6.10] Clark, G.C., and J.B. Cain, *Error-Correction Coding for Digital Communications,* Plenum Press, New York, 1981.

[7.1] Massey, J.L., *Threshold Decoding,* The MIT Press, Cambridge, MA, 1963.

[7.2] Reed, I.S., "A Class of Multiple-Error-Correcting Codes and Their Decoding Scheme," *IRE Trans. Information Theory,* Vol. PGIT-4, 1954, pp. 38–49.

[7.3] Robinson, J.P., "Error Propagation and Definite Decoding of Convolutional Codes," *IEEE Trans. Information Theory,* Vol. IT-14, January 1968, pp. 121–128.

[7.4] Massey, J.L., and M.K. Sain, "Inverses of Linear Sequential Circuits," *IEEE Trans. Computers,* Vol. C-17, No. 4, April 1968, pp. 330–337.

[8.1] Viterbi, A.J., "Error Bounds for Convolutional Codes and an Asymptotically Optimum Decoding Algorithm," *IEEE Trans. Information Theory,* Vol. IT-13, No. 2, 1967, pp. 260–269.

[8.2] Chase, D., M.M. Goutmann, and J.S. Zaborowski, "Maximum Likelihood (Viterbi) Decoding of Convolutional Codes," General Atronics Corp. Report 2054-98-586, January 1971.

[9.1] Wozencraft, J.M., *Sequential Decoding for Reliable Communication,* Ph.D. thesis and Research Laboratory of Electronics Report No. 325, Massachusetts Institute of Technology, Cambridge, MA, 1957.

[9.2] Reiffen, B., "Sequential Decoding for Discrete Memoryless Channels," *IRE Trans. Information Theory,* Vol. IT-8, No. 2, 1962, pp. 208–220.

[9.3] Fano, R.N., "A Heuristic Discussion of Probabilistic Decoding," *IRE Trans Information Theory,* Vol. IT-9, No. 2, 1963, pp. 64–74.

[9.4] Jelinek, F., "Fast Sequential Decoding Algorithm Using a Stack," *IBM J. Research and Development,* Vol. 13, 1969, pp. 675–685.

[9.5] Zigongirov, K. Sb., "Some Sequential Decoding Procedures," *Problemy Peredachi Informatsii,* Vol. I, No. 4, 1966, pp. 13–25. (English translation, *Problems of Information Transmission,* available through Plenum Publishing Corporation.)

[9.6] Reiffen, B., and D. Wiggert, "Some Experimental Results for a Sequential Decoding Algorithm Applied to an Asymmetric Channel," *M.I.T. Lincoln Laboratory Group Report 25G-17,* 1963.

[10.1] Benice, R.J., and A.H. Frey, Jr., "Comparisons of Error Control Techniques," *Tenth National Communications Symp.,* Utica, NY, 1964.

[10.2] Burton, H.O., and D.D. Sullivan, "Errors and Error Control," *Proc. IEEE,* Vol. 60, No. 11, November 1972, pp. 1293–1301.

[10.3] Fontaine, A.B., and R.G. Gallager, "Error Statistics and Coding for Binary Transmission over Telephone Circuits," *Proc. IRE,* Vol. 49, No. 6, June 1961, pp. 1059–1065.

[10.4] Reiffen, B., W.G. Schmidt, and H.L. Yudkin, "The Design of an Error Free Data Transmission System for Telephone Circuits," *Trans. AIEE* (*Communications and Electronics*), Vol. 80, 1961, pp. 224–231.

[10.5] Fire, P., "A Class of Multiple-Error-Correcting Binary Codes for Non-Independent Errors," *Report RSL-E-2,* Sylvania Reconnaissance Systems Laboratory, Mountain View, CA, 1959.

[10.6] Burton, H.O., "Inversionless Decoding of Binary BCH Codes," *IEEE Trans. Information Theory,* Vol. IT-17, No. 3, July 1971, pp. 464–466.

[10.7] Kohlenberg, A., and G.D. Forney, Jr., "Convolutional Coding for Channels with Memory," *IEEE Trans. Information Theory,* Vol. IT-14, No. 5, September 1968, pp. 618–626.

[10.8] Chase, D., "Code Combining a Maximum-Likelihood Decoding Approach for Combining an Arbitrary Number of Noisy Packets," *IEEE Trans. Communications,* Vol. COM-23, May 1985, pp. 385–393.

[10.9] Hallen, S., and D. Haccoun, "Sequential Decoding with ARQ and Code Combining, *MILCOM '87 Proc.,* Paper 23.1.

[11.1] Heller, J., and I.M. Jacobs, "Viterbi Decoding for Satellite and Space Communications," *IEEE Trans. Communications Technology,* Vol. COM-19, No. 5, Part 2, October 1971, pp. 835–848.

[12.1] Holmes, J.K., *Coherent Spread Spectrum Systems,* John Wiley and Sons, New York, 1982.

[12.2] Torrieri, D.D., *Principles of Secure Communications,* Artech House, Dedham, MA, 1985, Chapter 2.

[12.3] Barker, R.H., "Group Synchronizing of Binary Digital Systems," in *Communication Theory,* C. Cherry, ed., 1953, pp. 273–287.

[12.4] Pettit, R.L., *ECCM for Digital Communications,* Wadsworth, Belmont, CA, 1981, Chapters 3 and 4.

[12.5] Stiffler, J.J., *Theory of Synchronous Communications,* Prentice-Hall, Englewood Cliffs, NJ, 1971.

[12.6] Stiffler, J.J., "Synchronization of Telemetry Codes," *IEEE Trans. Software Engineering,* Vol. SE-8, No. 2, June 1962, pp. 112–117.

[12.7] Bloedel, E.D., "Synchronization During Biased PCM Conditions," *RCA Review,* December 1966, pp. 632–644.

Index

Printed in the United States
1279900001B/103